# PRACTICAL HETEROCYCLIC CHEMISTRY

# PRACTICAL HETEROCYCLIC CHEMISTRY

**A. O. Fitton and R. K. Smalley**
University of Salford, Lancs, England

**ACADEMIC PRESS**
London and New York
1968

ACADEMIC PRESS INC. (LONDON) LTD.
Berkeley Square House
Berkeley Square
London W.1.

U.S. Edition published by
ACADEMIC PRESS INC.
111 Fifth Avenue
New York, New York 10003



Library of Congress Catalog Card Number: 68–19255

PRINTED IN GREAT BRITAIN BY
SPOTTISWOODE, BALLANTYNE AND CO. LTD.
LONDON AND COLCHESTER

# INTRODUCTION

Recent publications concerned with laboratory work in organic chemistry have concentrated on semi-micro methods of preparation, and these tend to be confined to standard routes to aromatic and some aliphatic compounds. Chemistry remains a practical subject and a basic training should not be circumscribed by limiting factors of scale and type.

The authors emphasize the value of the exercises selected for inclusion as illustrative of the newer synthetic methods involving not only the recently available reagents but others that involve reactions proceeding through reactive intermediates. In order to achieve this special purpose the collection of systems may at first sight appear to be complex, but parent heterocyclic compounds are rarely available by direct synthesis; hence the inclusion of substituted compounds prepared from readily available intermediates.

The experiments have been carefully carried out by the authors, thereby checking the usefulness of the routes. The methods are especially suitable for completion within the time available for a normal laboratory class. By appropriate selection, the exercises can be made representative of requirements in college undergraduate courses at all levels and also in the recently introduced M.Sc. courses on aspects of advanced organic chemistry.

By providing this book the authors are meeting the need for a laboratory book making available a series of preparations as a supplement to lectures based on a modern approach to organic chemistry. The book is indeed the outcome of a need felt by these young lecturers to introduce into laboratory work a series of exercises that are typical of modern procedures and outlook and that form the basis of their lecture courses and research.

G. R. Ramage

# PREFACE

Heterocyclic chemistry has been the subject of many excellent factual and theoretical texts, but practical heterocyclic chemistry has received only sketchy attention in books dealing with general practical organic chemistry. In view of the commercial importance and widespread academic research interest in heterocyclic chemistry, the authors felt that a book dealing with practical aspects of this subject would help to familiarize students with the preparations and properties of heterocyclic compounds.

The book in fact contains one hundred heterocyclic preparations of varying complexity, and all have been checked by the authors. Examples will be found which are suitable for all years of an undergraduate course, and also for M.Sc. students. Since the procedures are usually general for a particular heterocycle, the book will also be of reference value to the research chemist.

Details of the actual preparations were obtained from many sources, and were selected so as to involve as wide a range of heterocyclic systems and reaction procedures as possible. Many of the important classical synthetic methods have been included, but a feature of the book are the examples that illustrate the use of newer synthetic methods. These are exemplified by reactions in which reactive intermediates, such as "carbenes", "nitrenes" and "benzynes", and reagents such as triphenylmethyl perchlorate, diketen and polyphosphoric acid are used. It should be made clear, however, that these intermediates and reagents apply to organic chemistry in general, and their use here merely illustrates their application to heterocyclic systems. Indeed many of the intermediate preparations involve the use of important classical reagents such as ethyl acetoacetate, malonate and diazonium salts, and are not themselves directly concerned with heterocyclic synthesis.

For reasons of consistency, some of the compound names used by the original authors have been altered, and the nomenclature adopted throughout the book is based on that used in publications of The Chemical Society.

The preparations vary in length from the single-stage reaction to the multi-stage or project-type synthesis, and where possible the duration of the various operations within an experiment has been limited so that the preparations can be carried out during a typical practical period of about three hours.

It is hoped that, where possible, the student will measure the spectra of the compound he has prepared, and where details of these have been reported, references have been included at the end of the preparation. The student can thus compare his results with those of earlier workers, or be assisted with the interpretation of his spectra. References are also given to the original authors of the preparations, and the "Reviews" sections deal with references to relevant material that will enable the student to read more about the synthesis or specific reagent used in the synthesis.

The authors thank Professor C. W. Rees (University of Leicester), Dr A. Gilbert (University of Reading), Professor H. Suschitzky and other colleagues at the University of Salford, for details of modern procedures that at the time of preparation of the manuscript had not appeared in the literature. They also wish particularly to thank Professor G. R. Ramage for contributing an Introduction, and finally Mr A. Rigby who carried out the tedious task of checking the references and percentage yields.

A. O. Fitton
R. K. Smalley

*Salford*
*February 1968*

# CONTENTS

# 1 FIVE-MEMBERED SYSTEMS CONTAINING ONE HETEROATOM

## 2,4-DIETHOXYCARBONYL-3,5-DIMETHYLPYRROLE

In this Knorr pyrrole synthesis,[1] excess of ethyl acetoacetate (I) is nitrosated, and the resulting nitroso compound is reduced by zinc and acetic acid. The reduced product reacts with unchanged ethyl acetoacetate yielding 2,4-diethoxycarbonyl-3,5-dimethylpyrrole (II).[2]

$$\underset{\text{I}}{Me\cdot CO\cdot CH_2\cdot CO_2Et} \xrightarrow{HNO_2} [Me\cdot CO\cdot CH(NO)\cdot CO_2Et] \xrightarrow{Zn/AcOH} [Me\cdot CO\cdot CH(NH_2)\cdot CO_2Et]$$

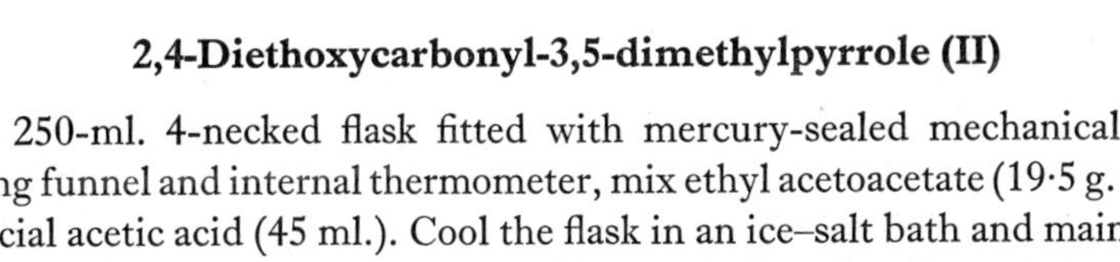

II

### 2,4-Diethoxycarbonyl-3,5-dimethylpyrrole (II)

In a 250-ml. 4-necked flask fitted with mercury-sealed mechanical stirrer, dropping funnel and internal thermometer, mix ethyl acetoacetate (19·5 g.; 19 ml.) and glacial acetic acid (45 ml.). Cool the flask in an ice–salt bath and maintain the internal temperature below 7° while adding a solution of sodium nitrite (5·35 g.) in water (9 ml.) with vigorous stirring. Remove the cooling-bath, and stir for a further 0·5 hr., during which time the flask contents attain room temperature.

Continue stirring, and add zinc dust (9·8 g.) portionwise over ca. 0·5 hr., and after adding further acetic acid (15 ml.) heat under reflux for 1 hr. Cool the solution, then decant from excess of zinc into stirred water (50 ml.). Allow the mixture to stand overnight, then filter off the crude product and wash with water.

2,4-Diethoxycarbonyl-3,5-dimethylpyrrole crystallizes from 95% aqueous ethanol (ca. 3 ml./g.) as buff needles, m.p. 138–139°.

Yield = 10–11 g. (56–62%).

## Spectra

(a) Ultraviolet.[2]
(b) Infrared.[3]

## Reviews

"Pyrroles", see review references on p. 3.

## References

1. L. Knorr and H. Lange, *Ber.*,1902, **35**, 2998.
2. A. Triebs and H. Kolm, *Annalen*, 1957, **606**, 166.
3. R. A. Jones and A. G. Moritz, *Spectrochim. Acta*, 1965, **21**, 295.

# 2,5-DIMETHYLPYRROLE

$\gamma$-Diketones are readily cyclized to 5-membered heterocycles[1] (see also p. 5). Acetonylacetone (I), in the presence of ammonium carbonate yields 2,5-dimethylpyrrole (II).[2]

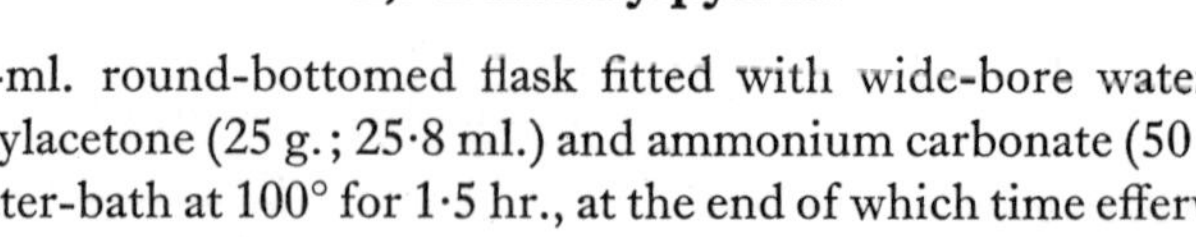


I II

## 2,5-Dimethylpyrrole

In a 250-ml. round-bottomed flask fitted with wide-bore water condenser, place acetonylacetone (25 g.; 25·8 ml.) and ammonium carbonate (50 g.). Heat the flask on a water-bath at 100° for 1·5 hr., at the end of which time effervescence will have ceased and a deposit of inorganic material will have collected in the condenser. Remove this deposit by rinsing the condenser down with water (ca. 10 ml.), then complete the reaction by gently heating under reflux for 0·5 hr.

Cool the resulting mixture, transfer into a separating funnel and separate the yellow-brown upper layer. Extract the aqueous layer with chloroform (10 ml.) and combine this extract with the organic layer. Dry ($MgSO_4$), and after evaporating off the chloroform, distil the product using an air condenser.

2,5-Dimethylpyrrole is obtained as a colourless liquid, b.p. 165–168°. Yield = 15–17·5 g. (70–82%).

### Note

2,5-Dimethylpyrrole darkens on exposure to oxygen and care should be taken to avoid this as far as possible. It should be stored under nitrogen in a dark bottle.

### Spectra

(a) Ultraviolet.[3]
(b) Infrared.[4]
(c) Proton magnetic resonance.[3]

### Reviews

(a) "The Chemistry of Pyrrole and its Derivatives", "Heterocyclic Compounds", ed. R. C. Elderfield, Vol. I, p. 277.
(b) "Pyrroles", "Chemistry of Carbon Compounds", ed. E. H. Rodd, Vol. IV A, p. 28.

### References

1. L. Lederer and C. Paal, *Ber.*, 1885, **18**, 2591.
2. D. M. Young and C. F. H. Allen, *Org. Synth.*, 1943, Coll. Vol. II, 219.
3. E. B. Whipple, Y. Chiang, and R. L. Hinman, *J. Amer. Chem. Soc.*, 1963, **85**, 26.
4. R. A. Jones and A. G. Moritz, *Spectrochim. Acta*, 1965, **21**, 295.

## 4,5-DIPHENYL-2-METHYLFURAN-3-CARBOXYLIC ACID

Acyloins react with diketen (I) yielding reduced furan-2-one derivatives,[1] which undergo re-arrangement in acid conditions to produce furan-3-carboxylic acids.[2]

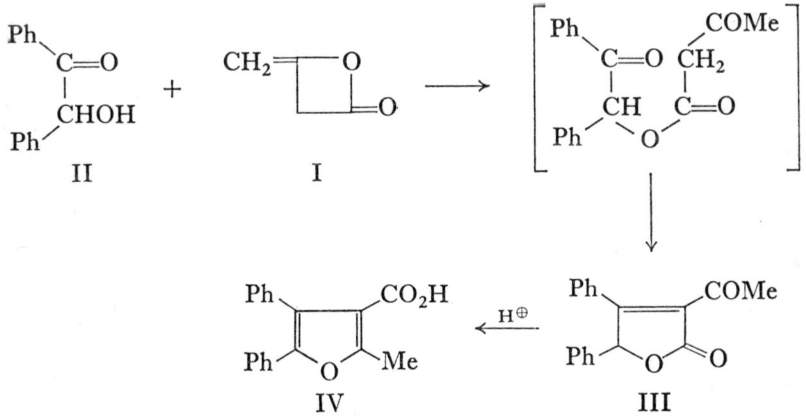

Use of benzoin (II) leads to 3-acetyl-2,5-dihydro-4,5-diphenylfuran-2-one (III),[1] which on re-arrangement gives 4,5-diphenyl-2-methylfuran-3-carboxylic acid (IV).[2]

### 3-Acetyl-2,5-dihydro-4,5-diphenylfuran-2-one (III)

In a 100-ml. round-bottomed flask, suspend benzoin (17·6 g.) in dry toluene (30 ml.) and add triethylamine (0·25 ml.) with shaking. To this mixture, add diketen† (7·5 g.; 6·9 ml.) and heat the flask *gently*, using a Dean and Stark apparatus[3] and reflux condenser, until an exothermic reaction occurs, then remove the source of heat, and allow the flask contents to bubble gently for ca. 10 min. When the reaction subsides, heat under reflux for a further 2 hr., when water (ca. 1 ml.) separates. Remove the solvent under reduced pressure to obtain the crude product as a brown oil that readily solidifies.

3-Acetyl-2,5-dihydro-4,5-diphenylfuran-2-one crystallizes from isopropanol (ca. 3 ml./g.) as off-white needles, m.p. 97–99°.

Yield = 14–16 g. (61–69%).

### 4,5-Diphenyl-2-methylfuran-3-carboxylic acid (IV)

In a 100-ml. conical flask, heat a solution of 3-acetyl-2,5-dihydro-4,5-diphenylfuran-2-one (5 g.) in a mixture of glacial acetic acid (50 ml.) and concentrated hydrochloric acid (5 ml.) on a water-bath at 100° for 1 hr. Cool, filter off the crude product, and wash with acetic acid.

4,5-Diphenyl-2-methylfuran-3-carboxylic acid crystallizes from ethyl acetate (ca. 10 ml./g.) as white hairy needles, m.p. 211–212°.

Yield = 2–2·2 g. (40–44%).

### Reviews

(a) "Furan", "Heterocyclic Compounds", ed. R. C. Elderfield, Vol. I, p. 119.
(b) "Furan and its Substitution Products", "Chemistry of Carbon Compounds", ed. E. H. Rodd, Vol. IV A, p. 138.
(c) "Some Aspects of Furan and Pyran Chemistry", D. G. Jones and A. W. C. Taylor, *Quart. Rev.*, 1950, **4**, 195.
(d) "Diketen", see review reference on p. 97.

### References

1. R. N. Lacey, *J. Chem. Soc.*, 1954, 816.
2. R. N. Lacey, *J. Chem. Soc.*, 1954, 822.
3. A.I. Vogel, "A Text-book of Practical Organic Chemistry", Longmans, Green & Co., London, 3rd. edn., 1956, p. 429.

† Diketen is an unstable highly lachrymatory liquid that should be stored at 0 ± 5°, and always handled carefully, using goggles and rubber gloves. Reactions involving diketen should be carried out in an efficient fume-cupboard.

# 2,5-DIMETHYLTHIOPHEN

γ-Diketones are readily cyclized to 5-membered heterocycles,[1] (see also p. 2). Acetonylacetone (I), in the presence of phosphorus pentasulphide yields 2,5-dimethylthiophen (II).[2]

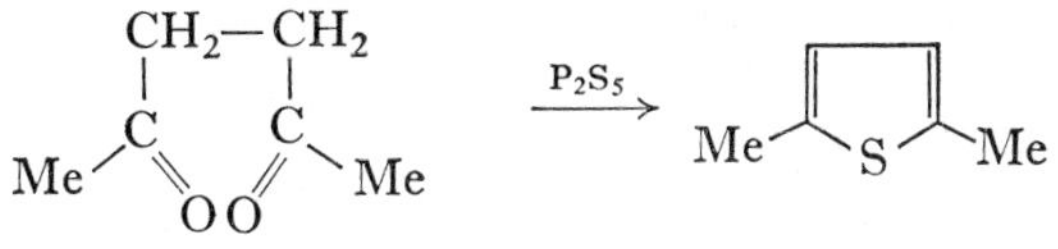

I II

## 2,5-Dimethylthiophen (II)

To phosphorus pentasulphide (11·1 g.) contained in a 100-ml. 2-necked round-bottomed flask fitted with reflux condenser and dropping funnel, add acetonylacetone (4·85 g.; 5 ml.), and heat the mixture *gently* until an exothermic reaction occurs. Remove the flame, and after the reaction has subsided, add the remainder of the acetonylacetone (18 g.; 18·5 ml.) from the dropping funnel at a rate such that the mixture boils gently. (Addition takes between 15 and 30 min.). Finally, complete the reaction by heating under reflux gently for a further 1 hr.

Cool and add the contents of the flask to ice–water (100 g.) and extract the mixture with ether (3 × 30 ml.). Combine the extracts, dry ($MgSO_4$), and after evaporating off the ether, distil the residual oil at atmospheric pressure.

2,5-Dimethylthiophen is obtained as a colourless, foul-smelling liquid, b.p. 134–136°.

Yield = 11–13 g. (49–58 %).

### Spectra

(a) Infrared.[3]
(b) Proton magnetic resonance.[4]

### Reviews

(a) "Thiophen and its Derivatives", "The Chemistry of Heterocyclic Compounds", ed. A. Weissberger, Vol. 3, p. 1.
(b) "The Chemistry of Thiophen", "Heterocyclic Compounds", ed. R. C. Elderfield, Vol. I, p. 208.
(c) "Thiophen and its Substitution Products", "Chemistry of Carbon Compounds", ed. E. H. Rodd, Vol. IV A, p. 203.

## References


1. C. Paal, *Ber.*, 1885, **18**, 2251.
2. M. W. Farrar and R. Levine, *J. Amer. Chem. Soc.*, 1950, **72**, 4433.
3. K. E. Schulte, A. Kreutzberger, and G. Bohn, *Chem. Ber.*, 1964, **97**, 3263.
4. G. S. Reddy and J. H. Goldstein, *J. Amer. Chem. Soc.*, 1961, **83**, 5020.

# 2 BENZODERIVATIVES OF FIVE-MEMBERED SYSTEMS CONTAINING ONE HETEROATOM

## CARBAZOLE

Azide decompositions normally proceed via an electron-deficient intermediate known as a nitrene,[1] or imidogen,[2] which can behave either as an electrophile or as a diradical. Smolinsky,[3] however, has indicated that the decomposition may proceed via a concerted process without the formation of a discrete intermediate. Such a mechanism is probably involved in the decomposition of 2-azidobiphenyl (II), which gives carbazole (III) in high yield.[4]

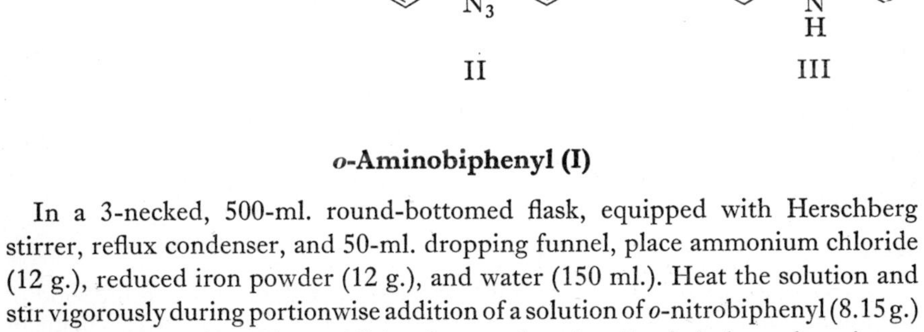


### *o*-Aminobiphenyl (I)

In a 3-necked, 500-ml. round-bottomed flask, equipped with Herschberg stirrer, reflux condenser, and 50-ml. dropping funnel, place ammonium chloride (12 g.), reduced iron powder (12 g.), and water (150 ml.). Heat the solution and stir vigorously during portionwise addition of a solution of *o*-nitrobiphenyl (8.15 g.) in ethanol (40 ml.). When addition is complete (ca. 5 min.), heat the mixture under reflux for 4 hr., during which time vigorous stirring should be continued. Transfer the reaction flask to a fume cupboard and steam-distil the black mixture.

**Note**

Compounds of this type tend to have carcinogenic properties, and therefore the amino-compound should *not* be steam-distilled in the open laboratory nor should it be handled carelessly.

After ca. 3 hr., filter the aqueous distillate and dissolve the residue in ether (100 ml.). Dry the ethereal solution ($MgSO_4$), and evaporate off the ether to obtain the product as a colourless oil that solidifies on cooling.

2-Aminobiphenyl is obtained as a white solid, m.p. 44–45°, sufficiently pure for direct use in the next stage.

Yield=4–5 g. (58–72%).

**Note**

A further 1–1·5 g. of amine can be obtained if the steam-distillation is continued for 6 hr.

## 2-Azidobiphenyl (II)

Dissolve 2-aminobiphenyl (2 g.) in a warm mixture of concentrated hydrochloric acid (18 ml.) and water (50 ml.). Cool the acid solution in an ice-bath, and add sodium nitrite (1·5 g.) dissolved in water (10 ml.), then stir the mixture at 0° for 5 min. Run the resulting clear diazonium solution into a solution of sodium acetate (60 g.) and sodium azide (2·5 g.) in water (100 ml.), contained in a 500-ml. conical flask, and allow to stir for 20 min.

**Note**

All reactions involving the use of sodium azide must be carried out in an efficient fume cupboard, since aqueous solutions of sodium azide release hydrazoic acid ($HN_3$) which is highly toxic.

Transfer the solution into a separating funnel and extract the semi-solid product with ether (2 × 70 ml.). Combine the extracts, and after washing with water (2 × 50 ml.), dry ($MgSO_4$) and evaporate off the ether on a water-bath at 60–70°, to obtain the product as a yellow oil that solidifies on cooling. (It is important not to heat the azide too strongly as decomposition may occur.)

2-Azidobiphenyl is obtained as pale yellow prisms, m.p. 49–50°, sufficiently pure for direct use in the next stage.

Yield=1·8–2·1 g. (78–91%).

**Note**

Nearly all azides decompose violently when heated alone either in the solid or the liquid state. The prepared azide must *not*, therefore, be dried over a flame or in an oven.

## Carbazole (III)

Dissolve 2-azidobiphenyl (2 g.) in 1-methylnaphthalene (50 ml.) and add the solution dropwise over a period of 10 min. to refluxing 1-methylnaphthalene (100 ml.) contained in a 250-ml. round-bottomed flask. Heat the solution under reflux for 20 min., then distil off the excess of 1-methylnaphthalene (ca. 140 ml.) by using water-pump vacuum. Cool and stir the brown residual oil with petroleum ether (b.p. 40–60°; 100 ml.) until a solid product is formed. Filter the mixture and wash the residue with petroleum ether (b.p. 40–60°; 2 × 30 ml.), and dry at 100°.

Carbazole crystallizes from xylene (ca. 12 ml./g.) as colourless plates, m.p. 237–238°.

Yield = 1·25–1·3 g. (73–76%).

### Spectra

(a) Ultraviolet.[5]
(b) Infrared.[6, 7]

### Reviews

(a) "Recent Advances in the Chemistry of Carbazoles", N. Campbell and B. M. Barclay, *Chem. Rev.*, 1947, **40**, 359.
(b) "Heterocyclic Compounds with Indole and Carbazole Systems", "The Chemistry of Heterocyclic Compounds", ed. A. Weissberger, Vol. 8, p. 70.
(c) "The Chemistry of Carbazole", "Heterocyclic Compounds", ed. R. C. Elderfield, Vol. 3, p. 291.
(d) "Carbazole", "Chemistry of Carbon Compounds", ed. E. H. Rodd, Vol. IV A, p. 120.
(e) "Alkyl and Aryl Azides", J. H. Boyer and F. C. Canter, *Chem. Rev.*, 1954, **54**, 1.
(f) For reviews on the chemistry of nitrenes and imidogens, see refs. 1 and 2.

### References

1. L. Horner and A. Christmann, *Angew. Chem.*, (*Internat. edn.*), 1963, **2**, 599.
2. R. A. Abramovitch and B. A. Davis, *Chem. Rev.*, 1964, **64**, 149.
3. G. Smolinsky, *J. Amer. Chem. Soc.*, 1961, **83**, 2489.
4. P. A. S. Smith and B. B. Brown, *J. Amer. Chem. Soc.*, 1951, **73**, 2435.
5. W. A. Schroeder, P. E. Wilcox, K. N. Trueblood, and A. O. Dekker, *Analyt. Chem.*, 1951, **23**, 1740.
6. C. G. Cannon and G. B. B. M. Sutherland, *Spectrochim. Acta*, 1951, **4**, 373.
7. R. A. Russell and H. W. Thompson, *J. Chem. Soc.*, 1955, 483.

## 1,2,3,4-TETRAHYDROCARBAZOLE

The phenylhydrazone derivatives of acyclic ketones are readily cyclized to indoles in the presence of zinc chloride (Fischer indole synthesis; see p. 13).

Borsche showed that the phenylhydrazone (I) of cyclohexanone undergoes a similar reaction to yield 1,2,3,4-tetrahydrocarbazole (II).[1] Glacial acetic acid is used in this case as the solvent for phenylhydrazone formation and also as the cyclizing agent.[2]

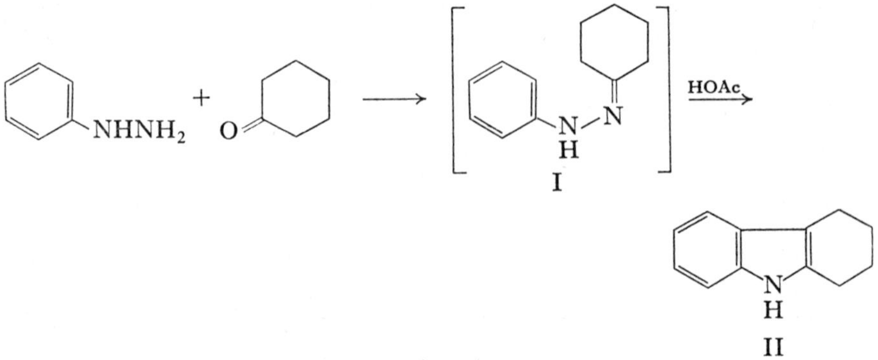


## 1,2,3,4-Tetrahydrocarbazole (II)

To a solution of cyclohexanone (4·75 g.; 5 ml.) in glacial acetic acid (30 ml.), contained in a 100-ml. round-bottomed flask, add phenylhydrazine (5·4 g.; 4·5 ml.). Heat the mixture under reflux for 10 min., then cool, and filter off the brown solid. Wash the product with cold water (4 × 50 ml.), and dry in a vacuum desiccator over concentrated sulphuric acid.

### Note

Tetrahydrocarbazole must not be dried in an oven, since it is readily oxidized to the hydroperoxide.

1,2,3,4-Tetrahydrocarbazole crystallizes from 50% aqueous methanol (ca. 6 ml./g.) (decolourizing carbon) as white prisms, m.p. 114–116°.

Yield = 6–7 g. (70–82%).

### Spectra

Ultraviolet.[3]

### Review

"Tetrahydrocarbazoles", "Chemistry of Carbon Compounds", ed. E. H. Rodd, Vol. IV A, p. 130.

### References

1. W. Borsche, A. Witte, and W. Bothe, *Annalen*, 1908, **359**, 49.
2. C. U. Rogers and B. B. Corson, *Org. Synth.*, 1963, Coll. Vol. IV, 884.
3. R. J. S. Beer, L. McGrath, and A. Robertson, *J. Chem. Soc.*, 1950, 2118.

## 2,3-DIOXOINDOLINE (ISATIN)

Cyclization of isonitrosoacetanilide (I) under strong acid conditions[1] (e.g. polyphosphoric acid, PPA) yields, 2,3-dioxoindoline (II). Isonitrosoacetanilide[2] is readily prepared from aniline, chloral hydrate, and hydroxylamine hydrochloride.

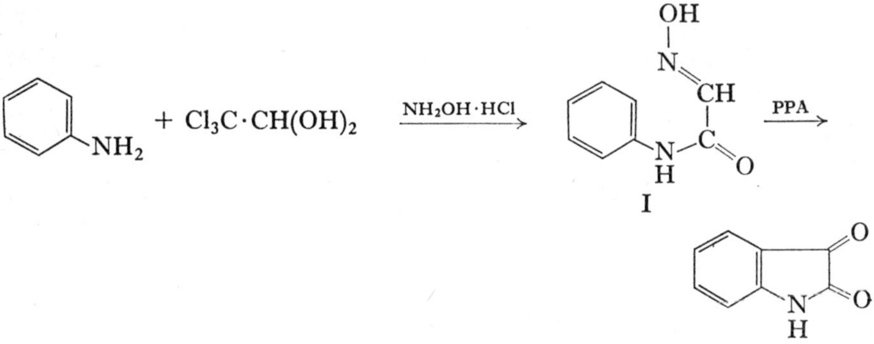


I

II

### Isonitrosoacetanilide (I)

Dissolve freshly distilled aniline (4·65 g.; 4·5 ml.) in concentrated hydrochloric acid (4·3 ml.) and water (30 ml.), and to the solution add a solution of chloral hydrate (9 g.) and sodium sulphate decahydrate (130 g.) in water (130 ml.). Next add a solution of hydroxylamine hydrochloride (11 g.) in water (150 ml.) and bring the resulting mixture to reflux temperature over a period of 15 min., then heat under reflux for 5 min. Allow the solution to cool and filter off the separated product. Wash the residue with cold water and dry at 100°.

Isonitrosoacetanilide is obtained as buff plates, m.p. 178–179°, sufficiently pure for direct use in the next stage.

Yield = 6·5–7·5 g. (79–91·5%).

### 2,3-Dioxoindoline (II)

To isonitrosoacetanilide (1 g.), add polyphosphoric acid (20 g.) and stir the mixture at 56° for 6 hr. (see apparatus on p. 38 and use acetone as heating liquid). Pour the resulting mixture into ice–water (100 ml.), stir vigorously, then filter off the precipitated product and wash with cold water.

2,3-Dioxoindoline crystallizes from glacial acetic acid (ca. 8 ml./g.) as orange prisms, m.p. 197°.

Yield = 0·6–0·7 g. (67–78%).

For a discussion on the tautomerism of oxindoles see ref. 3.

## Spectra

(a) Ultraviolet.[4]

(b) Infrared.[5]

## Reviews

(a) "The Chemistry of Isatin", W. C. Sumpter, *Chem. Rev.*, 1944, **34**, 393.

(b) "Isatin", "Chemistry of Carbon Compounds", ed. E. H. Rodd, Vol. IV A, p. 104.

(c) "Polyphosphoric acid", see review references on p. 78.

## References

1. E. B. Mullock and H. Suschitzky, University of Salford, unpublished results.
2. C. S. Marvel and G. S. Hiers, *Org. Synth.*, 1932, Coll. Vol. I, 327.
3. A. R. Katritzky and J. M. Lagowski, *Adv. Heterocyclic Chem.*, 1963, **2**, 16.
4. E. Sawicki, T. R. Hauser, and T. W. Stanley, *Analyt. Chem.*, 1959, **31**, 2063.
5. D. G. O'Sullivan and P. W. Sadler, *J. Chem. Soc.*, 1956, 2202.

## INDIGOTIN (INDIGO)

Reaction of *o*-nitrobenzaldehyde with acetone in the presence of sodium hydroxide yields indigotin.[1] The reaction proceeds via the aldol (I) which undergoes an intramolecular oxidation–reduction process to give indigotin (II).

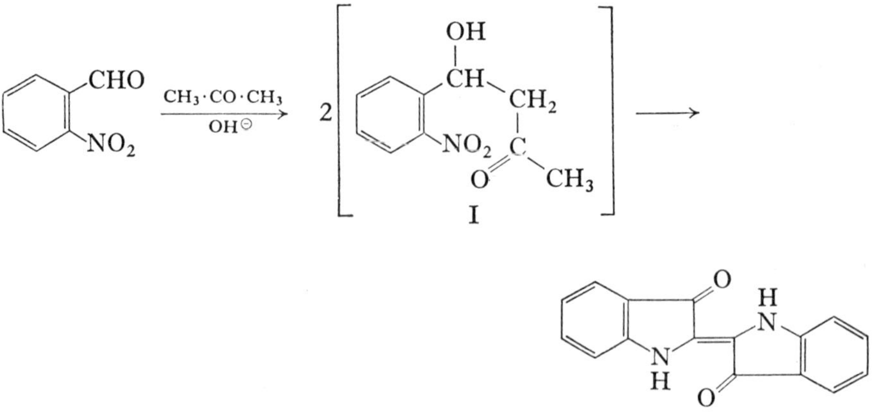


I

II

## Indigotin (II)

Dissolve *o*-nitrobenzaldehyde (1 g.) in a mixture of acetone (5 ml.) and water (5 ml.). Add dropwise 1N sodium hydroxide (5 ml.) and shake vigorously. The reaction mixture becomes quite warm and a colour change of green to brown should be noted. Allow the solution to stand for 5 min., then filter off the crystalline precipitate, and wash successively with ethanol ($3 \times 30$ ml.) and ether ($2 \times 20$ ml.).

Indigotin is obtained as dark-blue needles, m.p. $> 350°$, and can be crystallized from aniline.

Yield = 0·4–0·5 g. (46–57·5%).

### Spectra

(a) Ultraviolet.[2]
(b) Infrared.[3]

### Reviews

(a) "Heterocyclic Compounds with Indole and Carbazole Systems", "The Chemistry of Heterocyclic Compounds", ed. A. Weissberger, Vol. 8, p. 171.
(b) "Indigo (Indigotin) and its derivatives", "Chemistry of Carbon Compounds", ed. E. H. Rodd, Vol. IV B, p. 1083.

### References

1. A. von Baeyer and V. Drewson, *Ber.*, 1882, **15**, 2856.
2. P. W. Sadler, *J. Org. Chem.*, 1956, **21**, 317.
3. D. G. O'Sullivan, *J. Chem. Soc.*, 1960, 3278.

## 2-PHENYLINDOLE

Acetophenone phenylhydrazone (I) readily undergoes the Fischer indole[1] reaction to give 2-phenylindole (II). Polyphosphoric acid (PPA) can be used[2] advantageously as the cyclizing agent in place of zinc chloride.

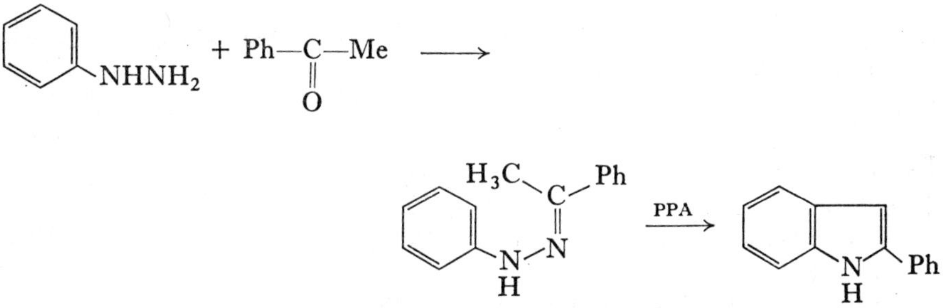

I II

## Acetophenone phenylhydrazone (I)

In a 50-ml. round-bottomed flask, place acetophenone (6 g.; 5·6 ml.) and freshly distilled phenylhydrazine (5·4 g.; 5 ml.). Heat the mixture on a water-bath at 100° for 1 hr., then cool and stir until the reaction mixture solidifies. Stir with ethanol (10 ml.), then filter off the product, and wash the residue with ice-cold ethanol (10 ml.).

Acetophenone phenylhydrazone is obtained as a yellow crystalline solid, m.p. 105°, sufficiently pure for direct use in the next stage.

Yield=9·5–10 g. (90·5–95%).

## 2-Phenylindole (II)

Add acetophenone phenylhydrazone (5 g.) to preheated (80°) polyphosphoric acid (50 g.), and stir the mixture at 100° for 1 hr. (see apparatus on p. 38 and use water as the heating liquid). Cool the solution, stir into ice–water (200 g.), and filter off the grey precipitate. Wash the residue with cold water (2×100 ml.), and dry at 100°.

2-Phenylindole crystallizes from ethanol (ca. 4 ml./g.) (decolourizing carbon) as white plates, m.p. 185–186°.

Yield=2·7–3 g. (59–65%).

### Spectra

(a) Ultraviolet.[3]
(b) Infrared.[4]

### Reviews

(a) "The Fischer-Indole Synthesis", P. A. Roussel, *J. Chem. Educ.*, 1953, **30**, 122.
(b) "The Fischer-Indole Synthesis", B. Robinson, *Chem. Rev.*, 1963, **63**, 373.
(c) "Indole", H. G. Lindwall and R. B. van Order, *Chem. Rev.*, 1942, **30**, 69.
(d) "Heterocyclic Compounds with Indole and Carbazole Systems", "The Chemistry of Heterocyclic Compounds", ed. A. Weissberger, Vol. 8, p. 1.
(e) "The Chemistry of Indoles", "Heterocyclic Compounds", ed. R. C. Elderfield, Vol. 3, p. 1.
(f) "Indoles and Hydro-indoles", "Chemistry of Carbon Compounds", ed. E. H. Rodd, Vol. IV A, p. 71.
(g) "Polyphosphoric acid", see review references on p. 78.

### References

1. E. Fischer and F. Jourdan, *Ber.*, 1883, **16**, 2241.
2. H. M. Kissman, D. W. Farnsworth, and B. Witkop, *J. Amer. Chem. Soc.*, 1952, **74**, 3948.
3. M. J. Kamlet and J. C. Dacons, *J. Org. Chem.*, 1961, **26**, 220.
4. R. A. Russell and H. W. Thompson, *J. Chem. Soc.*, 1955, 483.

## 2-BENZOYLBENZOFURAN (2-BENZOYLCOUMARONE)

Under basic conditions, *o*-hydroxyaldehydes condense with α-haloketones to yield benzofurans.[1] Interaction of salicylaldehyde with phenacyl bromide in the presence of potassium carbonate yields 2-benzoylbenzofuran (I).[2]

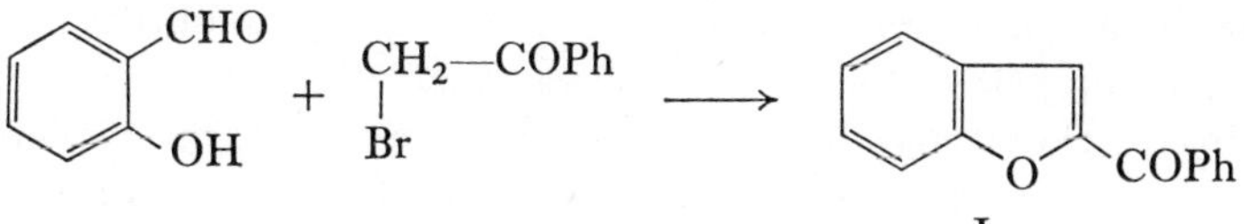


I

### 2-Benzoylbenzofuran (I)

To a solution of phenacyl bromide† (10 g.) and salicylaldehyde (6·1 g.; 5·2 ml.) in ethanol (10 ml.) contained in a 100-ml. round-bottomed flask, carefully add anhydrous potassium carbonate (6·9 g.). Heat the mixture on a water-bath at 100° for 1·5 hr., then stir the resulting orange suspension into water (150 ml.). After grinding up the precipitated solid, filter off and wash with water.

2-Benzoylbenzofuran crystallizes from ethanol (ca. 3 ml./g.) as elongated prisms, m.p. 89–91°.

Yield = 8·5–10 g. (77–90%).

**Reviews**

(a) "Benzofuran and its Derivatives", "Heterocyclic Compounds", ed. R. C. Elderfield, Vol. 2, p. 1.
(b) "Benzofurans", "Chemistry of Carbon Compounds", ed. E. H. Rodd, Vol. IV A, p. 168.

**References**

1. R. Stoermer and M. Schaffer, *Ber.*, 1903, **36**, 2863.
2. K. Freudenberg, H. Fikentscher, and M. Harder, *Annalen*, 1925, **441**, 157.

## 5-HYDROXYCYCLO-OCTANO[*b*]COUMARAN

In the presence of acid, cyclohexa-2,5-diene-4-one-spiro-2′-oxetano[3′,4′]-cyclo-octane (I) (p. 130) undergoes a dienone–phenol re-arrangement,[1] yielding 5-hydroxycyclo-octano[*b*]coumaran (II).[2]

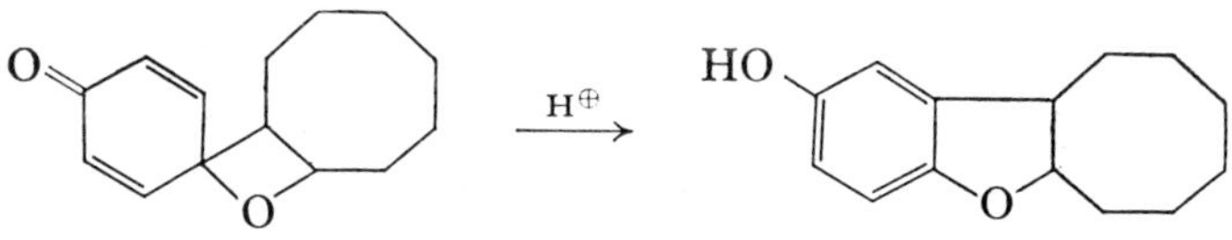


I II

† Phenacyl bromide is a powerful lachrymator. All operations involving its use should be carried out in an efficient fume cupboard.

### 5-Hydroxycyclo-octano[*b*]coumaran (II)

To a solution of cyclohexa-2,5-diene-4-one-spiro-2′-oxetano[3′,4′]cyclo-octane (1·25 g.) in ethanol (100 ml.), add concentrated hydrochloric acid (5 drops). Allow the resulting golden-brown solution to stand for 0·5 hr. at room temperature, then pour into water (400 ml.) and extract the mixture with ether (3 × 50 ml.). Combine the extracts, dry ($MgSO_4$) and after evaporating off the ether, distil the residual oil (ca. 1·2 g.) with oil-pump vacuum.

5-Hydroxycyclo-octano[*b*]coumaran is obtained as an oil that does not readily solidify. Slow evaporation of a benzene solution of the oil yields a crystalline product that recrystallizes from benzene (ca. 10 ml./g.) as colourless needles, m.p. 150–152°.

The yield depends on the efficiency of the distillation.

**Spectra**

(a) Ultraviolet.[2]
(b) Infrared.[2]
(c) Proton magnetic resonance.[2]

**Reviews**

"Coumarans", see review references on p. 17.

**References**

1. C. J. Collins and J. F. Eastham, "The Chemistry of the Carbonyl Group", ed. S. Patai, Interscience Publishers, Inc., New York, 1966, p. 775.
2. D. Bryce-Smith, A. Gilbert, and M. G. Johnson, *J. Chem. Soc.*, 1967, 383.

## 2-METHYLCOUMARAN

*o*-Allylphenols cyclize in acid conditions to yield either chromans or coumarans depending on the nature of the allyl side-chain.[1] *o*-Allylphenol (I) and hydrobromic acid yield 2-methylcoumaran (II).[2]

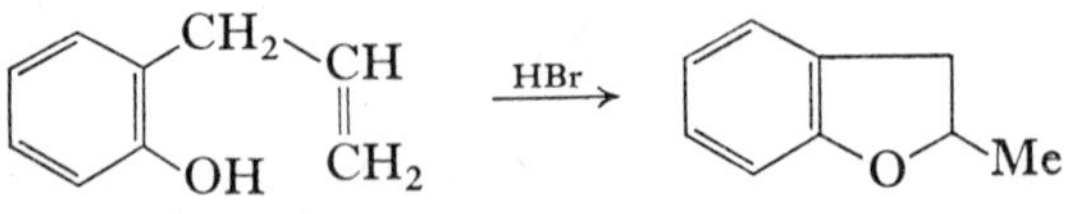


I II

### 2-Methylcoumaran (II)

In a 250-ml. round-bottomed flask, dissolve *o*-allylphenol (16·75 g.) in glacial acetic acid (70 ml.) and add 48% hydrobromic acid (35 ml.). Heat the solution under reflux for 0·5 hr., then cool and add water (100 ml.). Extract the liberated oil with ether (3 × 50 ml.), and wash the combined ethereal solution successively with water, 2N sodium hydroxide (3 × 25 ml.) and water. Dry the ethereal solution ($MgSO_4$), and after evaporating off the ether, distil the residual oil with water-pump vacuum.

2-Methylcoumaran is obtained as a colourless liquid, b.p. 85–88°/15 mm. Yield = 8–9 g. (48–54%).

### Spectra

Ultraviolet.[3]

### Reviews

(a) "Coumarans", "Heterocyclic Compounds", ed. R. C. Elderfield, Vol. 2, p. 43.
(b) "Coumarans (2,3-dihydrobenzofurans)", "Chemistry of Carbon Compounds", ed. E. H. Rodd, Vol. IV A, p. 170.

### References

1. L. I. Smith and J. A. King, *J. Amer. Chem. Soc.*, 1941, **63**, 1887.
2. L. Claisen, *Annalen*, 1919, **418**, 79, 83.
3. J. Entel, C. H. Ruofand, and H. C. Howard, *J. Amer. Chem. Soc.*, 1951, **73**, 4152.

## 5-CHLORO-3-METHYLBENZO[*b*]THIOPHEN

Cyclization of arylthioacetones can lead to a mixture of 2- and 3-substituted benzo[*b*]thiophens.[1,2] However, the composition of the product is affected by the reaction conditions.[2] Polyphosphoric acid (PPA) treatment of (*p*-chlorophenylthio)acetone (I), obtained from *p*-chlorothiophenol and chloroacetone leads to uncontaminated 5-chloro-3-methylbenzo[*b*]thiophen (II) in high yield.[2]

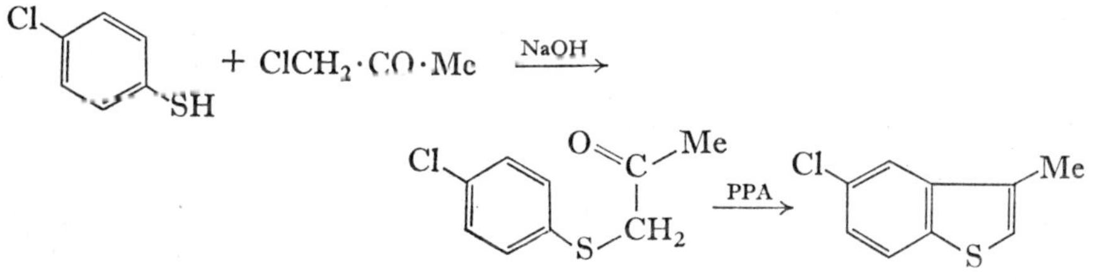


I II

### (*p*-Chlorophenylthio)acetone (I)

In a 250-ml. conical flask, prepare a solution of *p*-chlorothiophenol (18·26 g.) and sodium hydroxide (5 g.) in water (150 ml.). With magnetic stirring, add freshly distilled chloroacetone (11·63 g.; 10 ml.), then continue stirring at room temperature for 45 min. Transfer the mixture into a separating funnel and extract with ether (3 × 75 ml.). Combine the extracts, dry ($MgSO_4$) and after evaporating off the ether, distil the residual oil with water-pump vacuum.

(*p*-Chlorophenylthio)acetone is obtained as a pale-yellow oil, b.p. 160–164°/12 mm., which readily solidifies, m.p. 40–41°.

Yield = 18–21 g. (70–82%).

### 5-Chloro-3-methylbenzo[*b*]thiophen (II)

In a 250-ml. round-bottomed flask, mechanically stir a mixture of (*p*-chlorophenylthio)acetone (10 g.) and polyphosphoric acid (100 g.) at 125–130° for 45 min. Decompose the resulting complex by stirring into ice–water (300 ml.), then extract the mixture with ether (3 × 75 ml.). Combine the extracts, dry ($MgSO_4$) and after evaporating off the ether, distil the residual oil with water-pump vacuum.

5-Chloro-3-methylbenzo[*b*]thiophen is obtained as a colourless oil, b.p. 142–144°/14 mm., which readily solidifies, m.p. 32–34°.

Yield = 6·5–7·25 g. (71–79%).

**Reviews**

(a) "Thianaphthen and its Derivatives", "The Chemistry of Heterocyclic Compounds", ed. A. Weissberger, Vol. 7, p. 17.
(b) "Thionaphthen", "Heterocyclic Compounds", ed. R. C. Elderfield, Vol. 2, p. 145.
(c) "Benzothiophens (thionaphthens)", "Chemistry of Carbon Compounds", ed. E. H. Rodd, Vol. IV A, p. 225.
(d) "Polyphosphoric acid", see review references on p. 78.

**References**

1. D. S. Rao and B. D. Tilak, *J. Sci. Ind. Res., India*, 1959, **18B**, 77.
2. R. P. Dickinson and B. Iddon, University of Salford, unpublished results.

# 3 FIVE-MEMBERED SYSTEMS CONTAINING MORE THAN ONE HETEROATOM

## 2,4-DIOXOIMIDAZOLIDINE (HYDANTOIN)

2,4-Dioxoimidazolidines (hydantoins) result from acid treatment of α-ureido acid (hydantoic acid) derivatives (e.g. II) which are prepared from the corresponding α-amino-acid derivatives and potassium cyanate via the urea (Wöhler) re-arrangement. Ethyl aminoacetate hydrochloride (I) results from the Fischer–Speier[1] esterification of glycine, and is converted by potassium cyanate into ethyl ureidoacetate (II), which on treatment with concentrated hydrochloric acid affords 2,4-dioxoimidazolidine (III).[2]

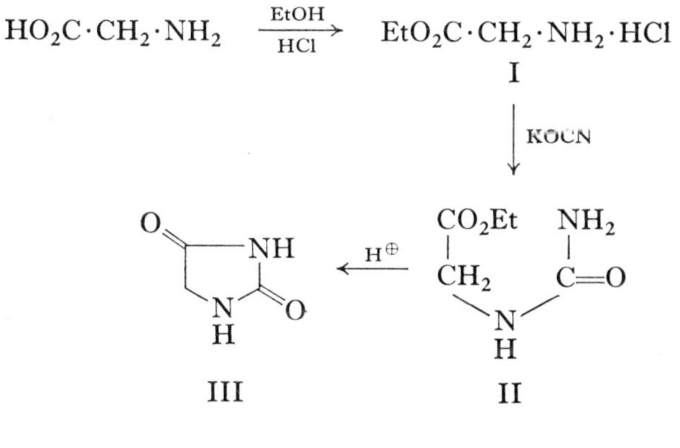


### Ethyl aminoacetate hydrochloride (I)

In a 100-ml. 2-necked round-bottomed flask, suspend glycine (7·5 g.) in ethanol (30 ml.). Heat the suspension under reflux on the water-bath, and with frequent swirling pass in dry hydrogen chloride gas (for method of generation, see p. 113) at the rate of 2–3 bubbles/sec., until solution is complete (ca. 30–45 min.). Allow the solution to cool, filter off the separated product, and wash with ethanol (15 ml.). Dry in a vacuum desiccator over sodium hydroxide pellets.

Ethyl aminoacetate hydrochloride is obtained as white needles, m.p. 144–145°, sufficiently pure for direct use in the next stage.

Yield=12–13 g. (86–93%).

### Ethyl ureidoacetate (II)

To a solution of ethyl aminoacetate hydrochloride (6·98 g.) in warm water (7 ml.), add a solution of potassium cyanate (4·05 g.) in water (8 ml.). Allow the resulting mixture to stand at room temperature for 20 min., then cool in an ice–water bath for 15 min. before filtering off the crude product.

Ethyl ureidoacetate crystallizes from ethanol (ca. 4 ml./g.) as colourless prisms, m.p. 134–135°.

Yield=5·5–6 g. (75–82%).

### 2,4-Dioxoimidazolidine (III)

In a 50-ml. round-bottomed flask heat a solution of ethyl ureidoacetate (5 g.) in concentrated hydrochloric acid (8 ml.) on the water-bath at 100° for 15 min. Evaporate the solution to dryness under reduced pressure to obtain the product as a white residue.

2,4-Dioxoimidazolidine crystallizes from ethanol (ca. 20 ml./g.) as white prisms, m.p. 220–222°.

Yield=2·4–3 g. (70–88%).

For a discussion on the tautomerism of hydantoins, see ref. 3.

### Spectra

(a) Ultraviolet.[4]
(b) Infrared.[5]

### Reviews

(a) "Hydantoins", "Heterocyclic Compounds", ed. R. C. Elderfield, Vol. 5, p. 254.
(b) "2,4-Diketoimidazolidines (Hydantoins)", "Chemistry of Carbon Compounds", ed. E. H. Rodd, Vol. IV A, p. 312.
(c) "The Chemistry of the Hydantoins", E. Ware, *Chem. Rev.*, 1950, **46**, 403.

### References

1. E. Fisher and A. Speier, *Ber.*, 1895, **28**, 3252.
2. C. Harries and M. Weiss, *Ber.*, 1900, **33**, 3418.
3. A. R. Katritzky and J. M. Lagowski, *Adv. Heterocyclic Chem.*, 1963, **2**, 54.
4. L. Crombie and K. C. Hooper, *J. Chem. Soc.*, 1955, 3010.
5. M. Horák and J. Gut, *Coll. Czech. Chem. Comm.*, 1961, **26**, 1680.

# 4,5-DIPHENYLIMIDAZOLE (4,5-DIPHENYLGLYOXALINE)

In the Radziszewski reaction,[1] α-dicarbonyl compounds condense with ammonia and aldehydes to form imidazoles. In this modified Radziszewski reaction, benzil, hexamine, and ammonium acetate react in acetic acid to yield 4,5-diphenylimidazole (I).[2]

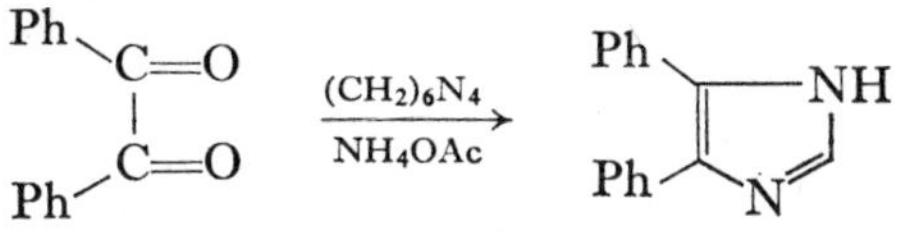

I

## 4,5-Diphenylimidazole (I)

In a 100-ml. round-bottomed flask, heat a solution of benzil (2·1 g.), hexamine (0·26 g.), and ammonium acetate ( 6 g.) in glacial acetic acid (50 ml.) under reflux for 1 hr. Add the resulting solution to water (400 ml.), and remove any turbidity that may develop by adding a small amount of carbon followed by stirring and filtration.

Precipitate the crude product by basifying the filtrate with concentrated ammonium hydroxide (70 ml.) with cooling, then filter and wash with water. Purify the residue by dissolving in hot pyridine (10 ml.) and carefully adding hot water (10 ml.). On cooling, 4,5-diphenylimidazole is obtained as fawn elongated prisms, m.p. 233–234°.

Yield = 1·7–1·9 g. (77–86%).

### Spectra

(a) Ultraviolet.[3]
(b) Infrared.[4]

### Reviews

(a) "Imidazole and its Derivatives", "The Chemistry of Heterocyclic Compounds", ed. A. Weissberger, Vol. 6.
(b) "Imidazoles and Condensed Imidazoles", "Heterocyclic Compounds", ed. R. C. Elderfield, Vol. 5, p. 194.
(c) "Imidazoles (glyoxalines)", "Chemistry of Carbon Compounds", ed. E. H. Rodd, Vol. IV A, p. 286.
(d) "Chemistry of Biologically Important Imidazoles", S. W. Fox, *Chem. Rev.*, 1943, **32**, 47.

**References**

1. B. Radziszewski, *Ber.*, 1882, **15**, 1493.
2. D. Davidson, M. Weiss, and M. Jelling, *J. Org. Chem.*, 1937, **2**, 328.
3. R. Gompper and H. Herlinger, *Chem. Ber.*, 1956, **89**, 2816.
4. D. M. White and J. Sonnenberg, *J. Org. Chem.*, 1964, **29**, 1926.

## 2-METHYL-Δ²-IMIDAZOLINE

Diacyl derivatives of ethylenediamine undergo cyclization to 2-substituted imidazolines. 1,2-Diacetamidoethane[1] (I) is cyclized by various reagents to 2-methyl-Δ²-imidazoline (II), the best cyclizing agent being magnesium.[2]

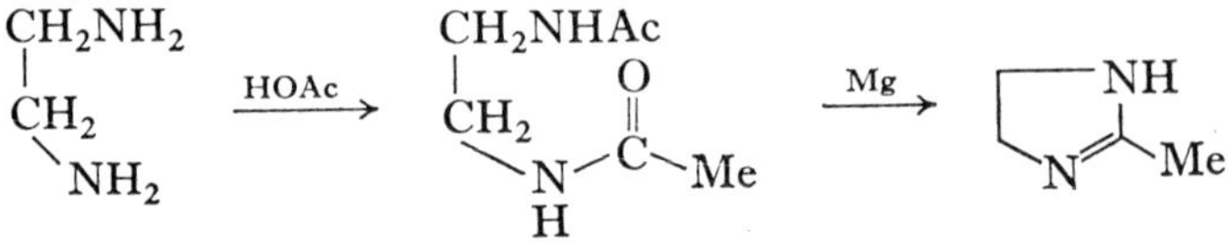


I II

### 1,2-Diacetamidoethane (I)

In a 100-ml. round-bottomed flask, cautiously mix ethylenediamine (15 g.; 16·7 ml.) with glacial acetic acid (30 g.; 28·6 ml.). Equip the flask with a still-head, thermometer, and condenser, and distil to remove water and excess of acid, keeping the still-head temperature below 115°. When all the volatile material (ca. 10 ml.) has been removed, cool the residual liquid with stirring when the crude product is obtained as a crystalline mass.

1,2-Diacetamidoethane crystallizes from ethanol (ca. 3 ml./g.) as white prisms, m.p. 172–174°.

Yield = 21–23 g. (58–64%).

### 2-Methyl-Δ²-imidazoline (II)

In a 50-ml. round-bottomed flask, place an intimately ground mixture of 1,2-diacetamidoethane (15 g.) and magnesium powder (2·5 g.). Equip the flask with a still-head, thermometer, and air condenser, and heat in a Wood's metal bath at 280–300° for 1·5 hr., collecting an off-white distillate that readily solidifies. Cool the distillation flask, and after breaking up the residue, extract this with hot benzene (100 ml.). Dissolve the solidified distillate in hot benzene (100 ml.), combine this solution with the benzene extract and stand the united solution over potassium hydroxide pellets (5 g.) for 15 min. Filter, and reduce the volume of the filtrate to ca. 15 ml., then allow to cool and filter off the crude product.

2-Methyl-$\Delta^2$-imidazoline crystallizes from benzene (ca. 3 ml./g.) (decolourizing carbon) as fine off-white needles, m.p. 93–95°.

Yield = 5–6 g. (57–69%).

## Spectra

(a) Ultraviolet.[3]
(b) Infrared.[4]

## Reviews

(a) "Imidazolines", "The Chemistry of Heterocyclic Compounds", ed. A. Weissberger, Vol. 6, p. 213.
(b) "Imidazolines", "Heterocyclic Compounds", ed. R. C. Elderfield, Vol. 5, p. 238.
(c) "Imidazolines", "Chemistry of Carbon Compounds", ed. E. H. Rodd, Vol. IV A, p. 301.
(d) "The Chemistry of 2-Imidazolines and Imidazolidines", R. J. Ferm and J. L. Riebsomer, *Chem. Rev.*, 1954, **54**, 593.

## References

1. L. H. Amundsen, *J. Chem. Educ.*, 1937, **14**, 141.
2. H. C. Chitwood and E. E. Reid, *J. Amer. Chem. Soc.*, 1935, **57**, 2424.
3. R. J. Ferm, J. L. Riebsomer, E. L. Martin, and G. H. Daub, *J. Org. Chem.*, 1953, **18**, 643.
4. P. Bassignana, C. Cogrossi, M. Gandino, and P. Merli, *Spectrochim. Acta*, 1965, **21**, 605.

# 3,5-DIMETHYLPYRAZOLE

Pyrazoles are readily prepared by the action of hydrazine or substituted hydrazines on 1,3-diketones.[1] The reaction proceeds via ring closure of the initially formed hydrazone. Interaction of acetylacetone with hydrazine hydrate leads to 3,5-dimethylpyrazole (I).[2]

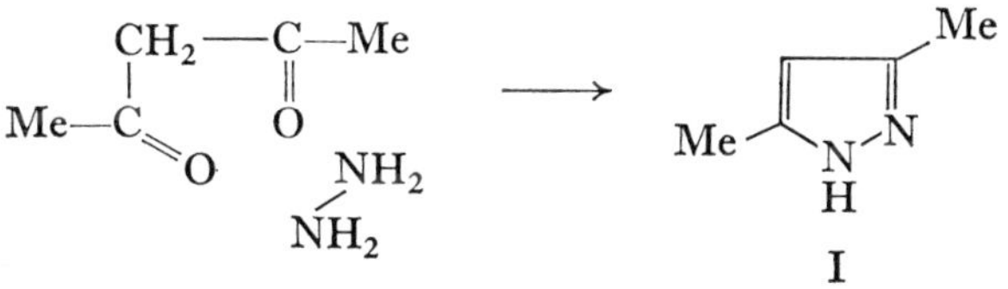


I

### 3,5-Dimethylpyrazole (I)

In a 100-ml. round-bottomed flask, dissolve acetylacetone (10 g.; 10·2 ml.), and hydrazine hydrate (5 ml.) in ethanol (50 ml.). Stir the mixture at room temperature for 10 min., and then heat under reflux for a further 0·5 hr. Pour the cooled solution into saturated brine (100 ml.), transfer into a separating funnel and extract the oily yellow product with ether (2 × 20 ml.). Combine the extracts, dry ($MgSO_4$), and evaporate off the ether to obtain the crude product as a yellow residue.

3,5-Dimethylpyrazole crystallizes from petroleum ether (b.p. 80–100°; ca. 6 ml./g.) as colourless plates, m.p. 106°.

Yield = 7–7·5 g. (73–78%).

## Spectra

(a) Ultraviolet.[3]
(b) Infrared.[4]
(c) Proton magnetic resonance.[5]

## Reviews

(a) "Pyrazoles and Related Compounds", "Heterocyclic Compounds", ed. R. C. Elderfield, Vol. 5, p. 45.
(b) "Pyrazoles", "Chemistry of Carbon Compounds", ed. E. H. Rodd, Vol. IV A, p. 244.

## References

1. E. Fischer and C. Bülow, *Ber.*, 1885, **18**, 2135.
2. P. E. Hexner and R. H. Wiley, *Org. Synth.*, 1963, Coll. Vol. IV, 351.
3. A. Mangini and D. D. M. Casoni, *Atti Accad. naz. Lincei, Rend. Classe Sci. fis. mat. nat.*, 1952, **13**, 46.
4. J. Charette and P. Teyssié, *Spectrochim. Acta*, 1959, **15**, 70.
5. I. L. Finar and E. F. Mooney, *Spectrochim. Acta*, 1964, **20**, 1269.

## 3-METHYL-1-PHENYL-5-PYRAZOLONE

Condensation of ethyl acetoacetate and its derivatives with various phenylhydrazines gives rise to the corresponding phenylhydrazones, which yield pyrazolones on cyclization. Interaction of ethyl acetoacetate and phenylhydrazine leads directly to 3-methyl-1-phenyl-5-pyrazolone (I).[1]

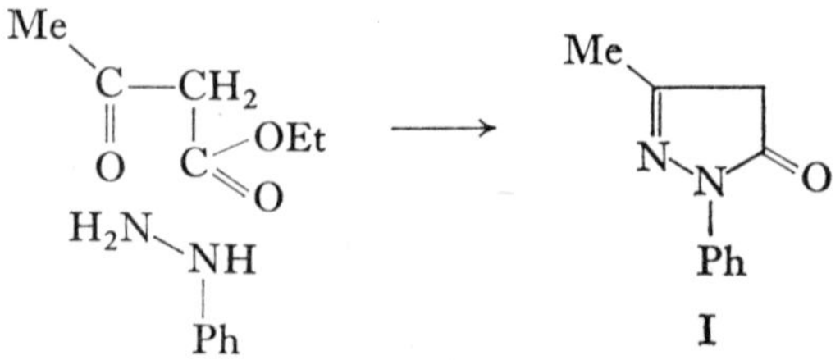


I

### 3-Methyl-1-phenyl-5-pyrazolone (I)

In a 100-ml. bolt-head flask, heat a mixture of ethyl acetoacetate (6·5 g.; 6·35 ml.) and freshly distilled phenylhydrazine (5·4 g.; 4·9 ml.) at 120° (oil-bath, temp. 120–130°) for 1 hr. Cool the resulting red oil and stir with ether (50 ml.) until solidification occurs, then filter off the crude product.

3-Methyl-1-phenyl-5-pyrazolone crystallizes from 50% aqueous ethanol (ca. 3 ml./g.) as colourless needles, m.p. 125–127°.

Yield = 6·5–7·5 g. (75–86%).

For a discussion on the tautomerism of pyrazolones, see ref. 2.

#### Spectra

(a) Ultraviolet.[3]
(b) Infrared.[4]
(c) Proton magnetic resonance.[4]

#### Reviews

(a) "Pyrazolin-5-ones", "The Chemistry of Heterocyclic Compounds", ed. A. Weissberger, Vol. 20, p. 13.
(b) "Hydroxypyrazoles and Pyrazolones", "Heterocyclic Compounds", ed. R. C. Elderfield, Vol. 5, p. 111.
(c) "Pyrazolones", "Chemistry of Carbon Compounds", ed. E. H. Rodd, Vol. IV A, p. 268.

#### References

1. L. Knorr, *Ber.*, 1883, **16**, 2597.
2. A. R. Katritzky and J. M. Lagowski, *Adv. Heterocyclic Chem.*, 1963, **2**, 40.
3. G. Westoo, *Acta Chem. Scand.*, 1952, **6**, 1499.
4. A. R. Katritzky and F. W. Maine, *Tetrahedron*, 1964, **20**, 299.

## 1,3,5-TRIPHENYL-$\Delta^2$-PYRAZOLINE

Nitrile imines behave as 1,3-dipoles, and as such will readily react with a variety of dipolarophiles to yield various heterocycles. For example, the nitrile imine obtained by treating α-chlorobenzaldehyde phenylhydrazone[1] (II) with base, smoothly adds across the double bond of styrene to yield 1,3,5-triphenyl-$\Delta^2$-pyrazoline (III).[2]

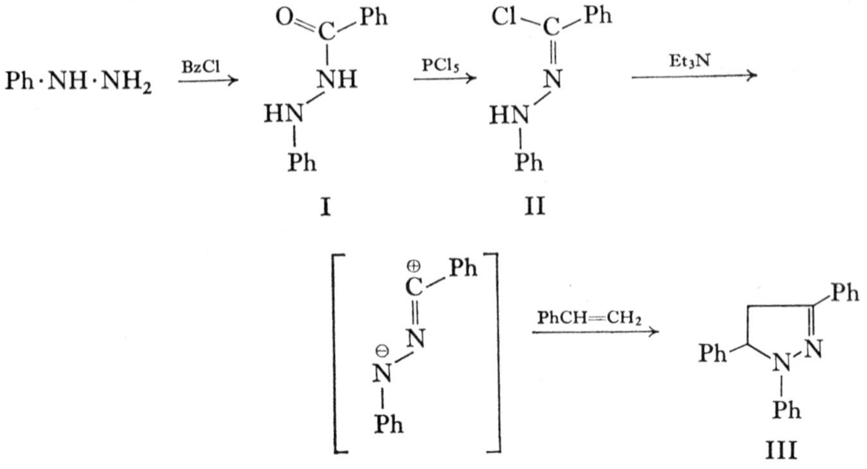


### β-Benzoylphenylhydrazine (I)

To a solution of freshly distilled phenylhydrazine (21·6 g.; 19·7 ml.) in pyridine (100 ml.) contained in a 250-ml. beaker, add slowly with stirring over ca. 2 min. benzoyl chloride (14 g.; 11·5 ml.). Stir the solution for 20 min., then pour into cold water (1 litre) and stir until a flocculent yellow precipitate is obtained. Filter off the crude product, wash with cold water ($4 \times 100$ ml.), and dry thoroughly at 100°.

β-Benzoylphenylhydrazine crystallizes from ethanol (ca. 12 ml./g.) as colourless prisms, m.p. 168–169°.

Yield = 18·5–19 g. (87–89·5%).

**Note**

It is essential that the benzoylhydrazine is absolutely dry for the next stage of the synthesis.

### α-Chlorobenzaldehyde phenylhydrazone (II)

To an intimately ground mixture of dry β-benzoylphenylhydrazine (14·8 g.) and phosphorus pentachloride (18 g.) in a 250-ml. round-bottomed flask, add sodium-dried ether (50 ml.). Equip the flask with an efficient reflux condenser fitted with drying tube ($CaCl_2$), and heat the mixture under reflux overnight. Cool, and to the mixture add a solution of phenol (30 g.) in sodium-dried ether (50 ml.) followed by methanol (40 ml.). When the exothermic reaction has subsided, evaporate off the solvent to about half bulk, and allow the remaining solution to cool in a refrigerator for several hours.

α-Chlorobenzaldehyde phenylhydrazone crystallizes from the reaction mixture as pale-yellow prisms, m.p. 130°.

Yield = 5·5–6·5 g. (34–40%).

**Note**

The authors found that a greater yield (9·3–9·5 g.) of product was obtained when the reaction mixture was allowed to stand in the refrigerator for several days.

## 1,3,5-Triphenyl-$\Delta^2$-pyrazoline (III)

To dry benzene (50 ml.), contained in a 250-ml. round-bottomed flask, add freshly distilled styrene (1·56 g.; 1·75 ml.), hydroquinone (0·2 g.), α-chlorobenzaldehyde phenylhydrazone (2·3 g.) and triethylamine (2·75 ml.). Heat the mixture under reflux on a water-bath at 100° for 2 hr. with occasional shaking. Filter the hot fluorescent solution in order to remove the precipitate of triethylamine hydrochloride, and evaporate off the solvent to obtain the crude product as an orange solid.

1,3,5-Triphenyl-$\Delta^2$-pyrazoline crystallizes from 90% aqueous methanol (ca. 4 ml./g.) as lemon needles, m.p. 138–139°.

Yield = 2·4–2·6 g. (81–87·5%).

**Note**

Triphenylpyrazolines exhibit an intense blue fluorescence when dissolved in organic solvents, and are used commercially as optical whiteners.[3]

**Spectra**

(a) Ultraviolet.[2]

(b) Proton magnetic resonance.[4]

**Reviews**

(a) "1,3-Dipolar Cycloadditions—Past and Future", R. Huisgen, *Angew. Chem.* (*Internat. edn.*), 1963, **2**, 565.

(b) "Pyrazoles and Related Compounds", "Heterocyclic Compounds", ed. R. C. Elderfield, Vol. 5, p. 45.

**References**

1. H. von Pechmann and L. Seeberger, *Ber.*, 1894, **27**, 2121.
2. R. Huisgen, M. Seidel, G. Wallbillich, and H. Knupfer, *Tetrahedron*, 1962, **17**, 3.
3. A. Wagner, C. W. Schellhammer, and S. Petersen, *Angew. Chem.* (*Internat. edn.*), 1966, **5**, 699.
4. A. Hassner and M. J. Michelson, *J. Org. Chem.*, 1962, **27**, 3974.

## 3,5-DIMETHYLISOXAZOLE

Isoxazoles may be prepared by treating 1,3-diketones with hydroxylamine hydrochloride.[1] The initially formed monoxime readily cyclizes to yield the isoxazole. Interaction of acetylacetone with hydroxylamine hydrochloride leads to 3,5-dimethylisoxazole (I).

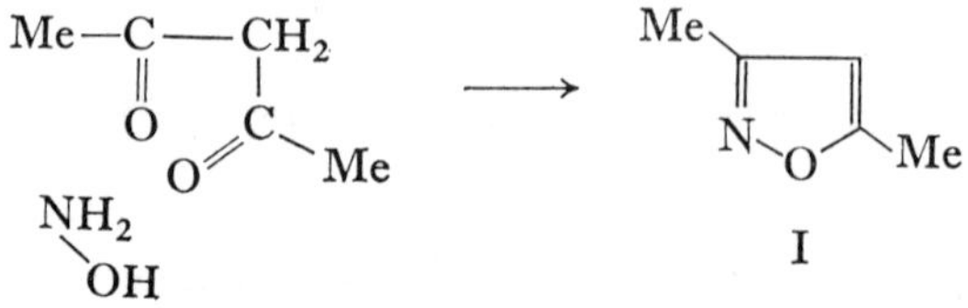

I

### 3,5-Dimethylisoxazole (I)

To a solution of hydroxylamine hydrochloride (7 g.) in water (15 ml.) contained in a 100-ml. round-bottomed flask, add a solution of acetylacetone (10 g.; 10·2 ml.) in ethanol (10 ml.). Heat the mixture under reflux until a negative ferric chloride test is given (ca. 1 hr.), and then pour the solution into cold water (60 ml.). Extract the aqueous mixture with ether (2 × 30 ml.), combine the extracts and dry ($MgSO_4$). Evaporate off the solvent, and transfer the brown oily residue to a 25 ml. pear-shaped flask. Distil, and neglect the first fraction (2–3 ml.; b.p. 80°), which is mainly ethanol.

3,5-Dimethylisoxazole is obtained as a colourless oil, b.p. 140–142°.

Yield = 5·5–6·5 g. (57–67%).

**Spectra**

(a) Ultraviolet.[2]

(b) Infrared.[3]

**Reviews**

(a) "Isoxazoles and Related Compounds", "The Chemistry of Heterocyclic Compounds", ed. A. Weissberger, Vol. 17, p. 1.

(b) "Isoxazoles", "Heterocyclic Compounds", ed. R. C. Elderfield, Vol. 5, p. 452.

(c) "Isoxazole, its Homologues and Substitution Products", "Chemistry of Carbon Compounds", ed. E. H. Rodd, Vol. IV A, p. 334.

(d) "Recent Developments in Isoxazole Chemistry", N. K. Kochetkov and S. D. Sokolov, *Adv. Heterocyclic Chem.*, 1963, **2**, 365.

**References**

1. L. Claisen, *Ber.*, 1891, **24**, 3900.
2. P. G. Dayton, *Compt. rend.*, 1953, **237**, 185.
3. A. R. Katritzky and A. J. Boulton, *Spectrochim. Acta*, 1961, **17**, 238.

## 3-PHENYL-5-ISOXAZOLONE

5-Isoxazolones are readily prepared by the action of hydroxylamine hydrochloride on β-keto-esters.[1] Initial reaction of the hydroxylamine with the ketone group yields an oxime (I), which then cyclizes to the 5-isoxazolone. Interaction of ethyl benzoylacetate with hydroxylamine hydrochloride yields 3-phenyl-5-isoxazolone (II).

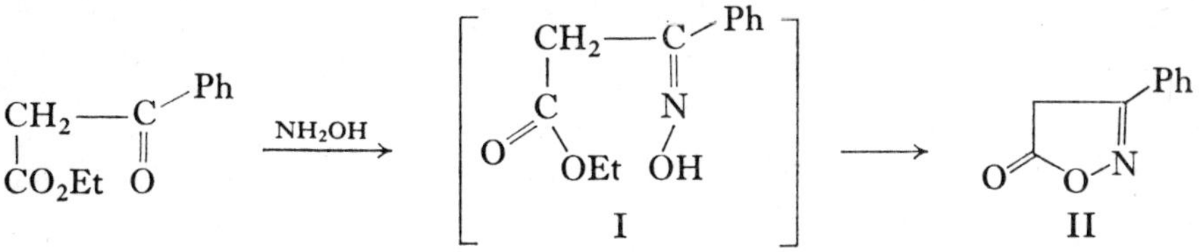


### 3-Phenyl-5-isoxazolone (II)

To a solution of ethyl benzoylacetate (9·6 g.; 8·5 ml.) in ethanol (20 ml.), contained in a 100-ml. round-bottomed flask, add a solution of hydroxylamine hydrochloride (3·5 g.) and sodium acetate (5 g.) in water (12 ml.). Heat the mixture under reflux for 2 hr., then allow to cool and filter off the crystalline product.

3-Phenyl-5-isoxazolone crystallizes from ethanol (ca. 12 ml./g.) as colourless plates, m.p. 152°.

Yield = 4–5 g. (50–62%).

For a discussion on the tautomerism of isoxazolones, see ref. 2.

### Spectra

(a) Ultraviolet.[3]
(b) Infrared.[3]
(c) Proton magnetic resonance.[4]

### Reviews

(a) "Isoxazoles and Related Compounds", "The Chemistry of Heterocyclic Compounds", ed. A. Weissberger, Vol. 17, p. 95.
(b) "Isoxazoles", "Heterocyclic Compounds", ed. R. C. Elderfield, Vol. 5, p. 472.
(c) "Isoxazolines and Isoxazolones", "Chemistry of Carbon Compounds", ed. E. H. Rodd, Vol. IV A, p. 343.

### References

1. A. Hantzsch, *Ber.*, 1891, **24**, 502.
2. A. R. Katritzky and J. M. Lagowski, *Adv. Heterocyclic Chem.*, 1963, **2**, 36.
3. A. R. Katritzky and A. J. Boulton, *Tetrahedron*, 1961, **12**, 41.
4. A. E. Hydorn, F. A. McGinn, J. R. Moetz, and J. Schwartz, *J. Org. Chem.*, 1962, **27**, 4305.

## 4-BENZYLIDENE-2-PHENYL-5-OXAZOLONE

Aldehydes readily condense with α-acylamino-acids in the presence of acetic anhydride to yield 2,5-disubstituted 5-oxazolones (azlactones).[1] The reaction probably involves prior cyclization of the acylated amino-acid to the 5-oxazolone (I), the reactive methylene group of which (position 4), then condenses with the aldehyde to yield the arylidene derivative (II).[2]

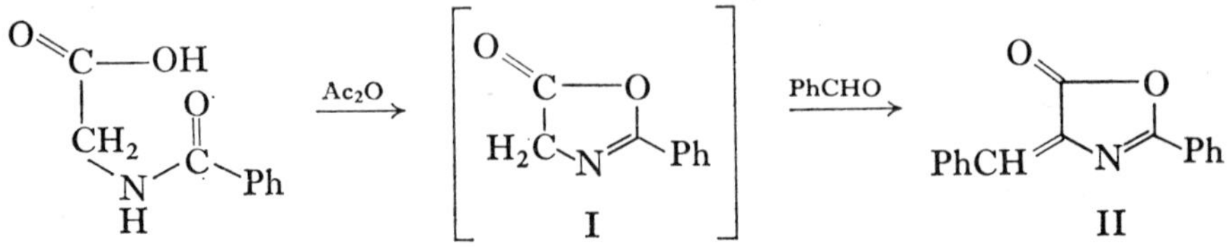


### 4-Benzylidene-2-phenyl-5-oxazolone (II)

Prepare an intimate mixture of benzaldehyde (3·3 g.; 3·1 ml.), hippuric acid (5·6 g.), anhydrous sodium acetate (2·6 g.), and acetic anhydride (20 ml.) in a 100-ml. conical flask, and warm carefully (avoid charring) over a Bunsen burner, until the reaction mixture becomes completely molten. Heat the bright-yellow liquid mixture on a water-bath at 100° for 1 hr., during which time some product may crystallize out. Cool the mixture, cautiously add ethanol (30 ml.), and warm on a water-bath for 0·5 hr., by which time all the acetic anhydride will have been converted to ethyl acetate. Cool the reaction mixture, filter off the yellow product, and wash successively with boiling water (2 × 50 ml.) and ice-cold ethanol (2 × 25 ml.).

4-Benzylidene-2-phenyl-5-oxazolone crystallizes from benzene (ca. 6 ml./g.) as bright-yellow needles, m.p. 166–167°.

Yield = 4–4·5 g. (51·5–57·5%).

### Spectra

(a) Ultraviolet.[3]
(b) Infrared.[4]

### Reviews

(a) "Azlactones", H. E. Carter, *Org. Reactions*, 1946, **3**, 198.
(b) "The Chemistry of 5-Oxazolones", E. Baltazzi, *Quart. Rev.*, 1955, **9**, 150.
(c) "Oxazole and its Derivatives", "Heterocyclic Compounds", ed. R. C. Elderfield, Vol. 5, p. 336.
(d) "5-Oxazolones", "Chemistry of Carbon Compounds", ed. E. H. Rodd, Vol. IV A, p. 365.

**References**

1. E. Erlenmeyer, *Annalen,* 1892, **271**, 137; 1904, **337**, 265.
2. J. S. Buck and W. S. Ide, *Org. Synth.*, 1943, Coll. Vol. II, 55.
3. D. A. Bassi, V. Deulofeu, and F. A. F. Ortega, *J. Amer. Chem. Soc.*, 1953, **75**, 171.
4. F. Micheel and B. Schleppinghoff, *Ber.*, 1955, **88**, 763.

## CYCLOHEXANESPIRO-2′-(1′,3′-OXAZOLIDINE)

Condensation of ketones with ethanolamine (I) can lead to alkylidene amino-alcohols or oxazolidines.[1] Cyclohexanone and ethanolamine yield cyclohexane-spiro-2′-(1′,3′-oxazolidine) (II).[1]

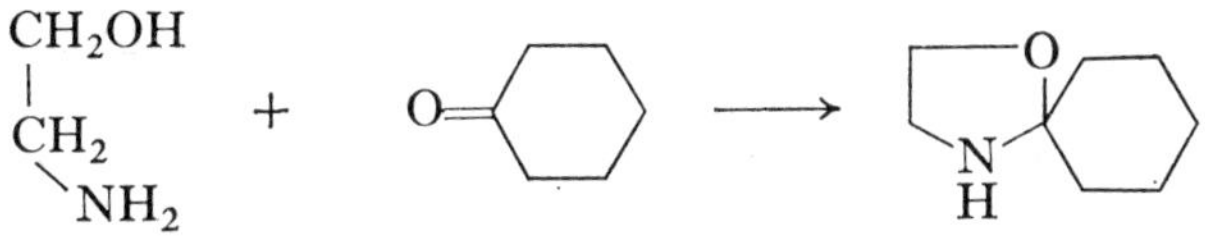


I II

### Cyclohexanespiro-2′-(1′,3′-oxazolidine) (II)

In a 100-ml. round-bottomed flask, dissolve ethanolamine (15·3 g.; 15 ml.) in benzene (50 ml.), and add cyclohexanone (32 g.; 34 ml.). Heat the flask for 0·5 hr. using a Dean and Stark apparatus[2] and reflux condenser, when water (ca. 4 ml.) separates. Evaporate off the benzene, and carefully fractionate the residue under water-pump vacuum.

Cyclohexanespiro-2′-(1′,3′-oxazolidine) is obtained as a colourless liquid, b.p. 89–91°/16 mm.

Yield = 27–33 g. (77–94%).

**Reviews**

(a) "Oxazolidines", "Heterocyclic Compounds", ed. R. C. Elderfield, Vol. 5, p. 391.
(b) "Oxazolidines", "Chemistry of Carbon Compounds", ed. E. H. Rodd, Vol. IV A, p. 372.
(c) "The Oxazolidines", E. D. Bergmann, *Chem. Rev.*, 1953, **53**, 309.

**References**

1. A. C. Cope and E. M. Hancock, *J. Amer. Chem. Soc.*, 1942, **64**, 1503.
2. A. I. Vogel, "A Text-book of Practical Organic Chemistry", Longmans, Green & Co., London, 3rd edn., 1956, p. 429.

## 2-METHYL-$\Delta^2$-OXAZOLINE

$\beta$-Hydroxyalkylamides yield $\Delta^2$-oxazolines on cyclodehydration in, for example, sulphuric acid. *N*-Acetylethanolamine (II), which is readily obtained[1] from ethanolamine (I) and acetic acid, is cyclized to 2-methyl-$\Delta^2$-oxazoline (III) merely by heat treatment.[1]

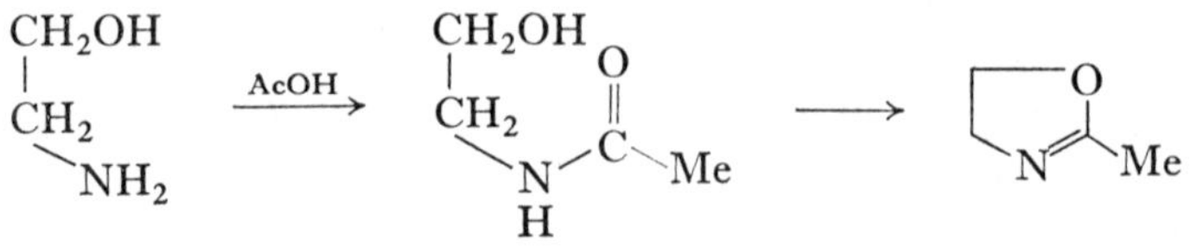


I II III

### *N*-Acetylethanolamine (II)

Cautiously mix ethanolamine (61 g.; 60 ml.) and glacial acetic acid (60 g.; 57 ml.) in a 250-ml. 2-necked, round-bottomed flask fitted with an internal thermometer. Equip the flask with a still-head and condenser etc., then heat gradually until water starts to distil (internal temperature ca. 165°). Continue heating until approximately 18 ml. of water have been collected when the internal temperature will have risen to ca. 205°, (time required is ca. 1 hr.). Distil the residual oil under oil-pump vacuum.

*N*-Acetylethanolamine is obtained as a colourless, viscous liquid, b.p. 141–149°/1·5 mm.

Yield = 85–90 g. (82·5–87·5%).

### 2-Methyl-$\Delta^2$-oxazoline (III)

In a 250-ml. 2-necked, round-bottomed flask fitted with internal thermometer and equipped for distillation, place *N*-acetylethanolamine (30·9 g.). Raise the internal temperature to 260° during 15 min., then increase to 310° over 1 hr., during which time the distillation temperature will increase from 105° to 150°, and ca. 16 g. of distillate will be obtained.

Transfer the distillate into a 250-ml. separating funnel, extract with ether (100 ml.) and run off the aqueous layer. Dry the ethereal solution ($MgSO_4$), and after evaporating off the ether distil the residual oil at atmospheric pressure.

2-Methyl-$\Delta^2$-oxazoline is obtained as a colourless, foul-smelling liquid, b.p. 108–111°.

Yield = 6·4–7·6 g. (25–30%).

**Spectra**

(a) Ultraviolet.[2]
(b) Infrared.[3]
(c) Proton magnetic resonance.[4]

**Reviews**

(a) "Oxazolines", "Heterocyclic Compounds", ed. R. C. Elderfield, Vol. 5, p. 377.
(b) "Oxazolines", "Chemistry of Carbon Compounds", ed. E. H. Rodd, Vol. IV A, p. 361.
(c) "The Chemistry of the Oxazolines", R. H. Wiley and L. L. Bennett jun., *Chem. Rev.*, 1949, **44**, 447.

**References**

1. H. Wenker, *J. Amer. Chem. Soc.*, 1935, **57**, 1079.
2. K. Kormendy and J. Volford, *Acta Chim. Acad. Sci. Hung.*, 1962, **32**, 115.
3. P. Bassignana, C. Cogrossi, and M. Gandino, *Spectrochim. Acta*, 1963, **19**, 1885.
4. M. A. Weinberger and R. Greenhalgh, *Canad. J. Chem.*, 1963, **41**, 1038.

## 2,4,5-TRIPHENYLOXAZOLE

On treatment with ammonium acetate in acetic acid, desyl benzoate (II), obtained from desyl alcohol (benzoin) (I) by benzoylation in pyridine, yields 2,4,5-triphenyloxazole (III).[1]

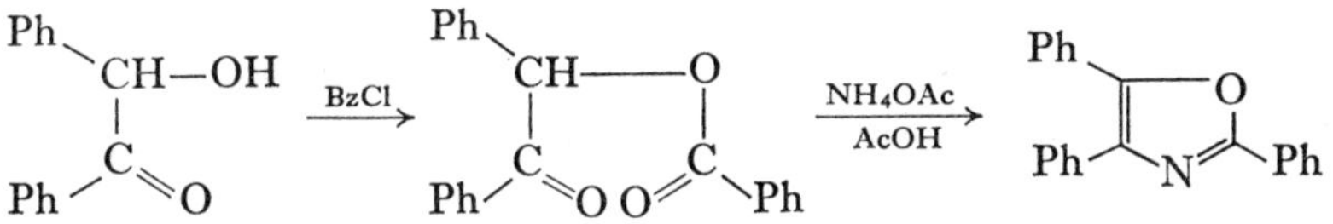


I II III

### Desyl benzoate (II)

To a solution of benzoin (5·3 g.) in pyridine (10 ml.) in a 50-ml. conical flask, cautiously add benzoyl chloride (3·9 g.; 3·2 ml.). Warm the mixture on a water-bath at 100° for 15 min., and finally heat over a flame to complete solution. Add the resulting solution to cold water (100 ml.) and stir vigorously until the precipitated oil solidifies. After grinding up the solid, filter off and wash with water.

Desyl benzoate crystallizes from ethanol (ca. 10 ml./g.) as colourless prisms, m.p. 127–128°.

Yield = 5·5–6·5 g. (70–82%).

### 2,4,5-Triphenyloxazole (III)

In a 100-ml. round-bottomed flask, dissolve desyl benzoate (3·16 g.) and ammonium acetate (6 g.) in acetic acid (50 ml.), and heat the solution under reflux for 1 hr. Cool the resulting solution, then add water (40 ml.) when an oil is precipitated that solidifies on chilling in ice–water. Filter off the solid and wash with 60% aqueous acetic acid (10 ml.).

2,4,5-Triphenyloxazole crystallizes from ethanol (ca. 8 ml./g.) as lustrous elongated prisms, m.p. 115–116°.

Yield = 2·3–2·5 g. (77–84%).

## Spectra

Ultraviolet.[2]

## Reviews

(a) "Oxazole and its Derivatives", "Heterocyclic Compounds", ed. R. C. Elderfield, Vol. 5, p. 298.
(b) "Oxazoles", "Chemistry of Carbon Compounds", ed. E. H. Rodd, Vol. IV A, p. 353.
(c) "The Chemistry of the Oxazoles", R. H. Wiley, *Chem. Rev.*, 1945, **37**, 309.

## References

1. D. Davidson, M. Weiss, and M. Jelling, *J. Org. Chem.*, 1937, **2**, 319.
2. D. G. Ott, F. N. Hayes, E. Hansbury, and V. N. Kerr, *J. Amer. Chem. Soc.*, 1957, **79**, 5448.

# *p*-CHLOROPHENYLSYDNONE

Sydnones are meso-ionic compounds[1] prepared by dehydration of *N*-nitrosoglycines with acetic anhydride.[2] Condensation of *p*-chloroaniline with ethyl chloroacetate leads to the glycine ester (I), which on hydrolysis and reaction with nitrous acid gives *N*-nitroso-*N*-(*p*-chlorophenyl)glycine (III).[3] Treatment of (III) with acetic anhydride yields *N*-(*p*-chlorophenyl)sydnone (IV).[3]

$$p\text{-Cl}\cdot C_6H_4\cdot NH_2 \xrightarrow{ClCH_2\cdot CO_2Et} p\text{-Cl}\cdot C_6H_4\cdot NH\cdot CH_2\cdot CO_2Et \xrightarrow{OH^{\ominus}}$$

I

$$p\text{-Cl}\cdot C_6H_4\cdot NH\cdot CH_2\cdot CO_2H$$

II

$$\downarrow HNO_2$$

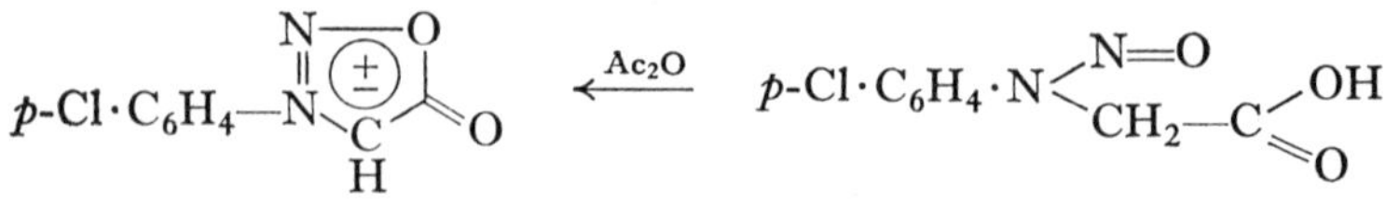

IV III

## *N*-(*p*-Chlorophenyl)glycine ethyl ester (I)

To a solution of ethyl chloroacetate (6·2 g.; 5·3 ml.) in ethanol (5 ml.), contained in a 100-ml. round-bottomed flask, add finely ground *p*-chloroaniline (6·4 g.) and hydrated sodium acetate (6 g.). Heat the mixture under reflux for 6 hr., then cool and stir into cold water (150 ml.). Filter off the pale yellow precipitate and wash with cold water (3 × 60 ml.).

*N*-(*p*-Chlorophenyl)glycine ethyl ester crystallizes from ethanol (ca. 5 ml./g.) as colourless plates, m.p. 88–90°.

Yield = 5 g.

A second crop of crystals (2–2·5 g.) may be obtained on adding water to the filtered ethanol solution.

Total yield = 7–7·5 g. (66–70·5%).

## *N*-(*p*-Chlorophenyl)glycine (II)

In a 100-ml. round-bottomed flask heat a mixture of *N*-(*p*-chlorophenyl)glycine ethyl ester (5 g.), sodium hydroxide (2 g.) and water (20 ml.) under reflux for 0·5 hr. Cool the resulting solution, then filter and acidify the filtrate with concentrated hydrochloric acid to pH 2. Filter off the precipitated product and dry in a vacuum desiccator.

*N*-(*p*-Chlorophenyl)glycine is obtained as a white solid, m.p. 140–141°, sufficiently pure for direct use in the next stage.

Yield = 3–3·5 g. (69–80·5%).

## *N*-Nitroso-*N*-(*p*-chlorophenyl)glycine (III)

To an ice-cold, stirred suspension of *N*-(*p*-chlorophenyl)glycine (3 g.) in water (30 ml.), contained in a 100 ml. conical flask, add slowly a solution of sodium nitrite (1·2 g.) in water (4 ml.), and stir the mixture, with ice–water cooling, until a clear yellow solution is obtained (ca. 2 hr.). Add concentrated hydrochloric acid (2 ml.), filter off the precipitated product and dry as far as possible by suction (ca. 0·5 hr.).

*N*-Nitroso-*N*-(*p*-chlorophenyl)glycine is obtained as pale green needles, m.p. 114°, sufficiently pure for direct use in the next stage.

### Note

The *N*-nitroso compound is unstable and, therefore, should not be dried in an oven. It is also advisable to carry out the next stage as soon as possible.

Yield = 3–3·5 g. (product still damp).

### *N*-(*p*-Chlorophenyl)sydnone (IV)

In a 100-ml. round-bottomed flask, heat a solution of *N*-nitroso-*N*-(*p*-chlorophenyl)glycine (3 g.) in acetic anhydride (20 ml.) on a water-bath at 100° for 1 hr. Evaporate the solution to dryness under reduced pressure, and purify the resulting crude product by crystallization from a mixture of benzene and petroleum ether (b.p. 60–80°) as follows. Dissolve the residue from evaporation in boiling benzene (30 ml.) and to the hot solution add hot petroleum ether (b.p. 60–80°) until a cloudiness occurs. At this point allow to cool and filter off the separated product.

*N*-(*p*-Chlorophenyl)sydnone is obtained as yellow needles, m.p. 112–113°.

Yield=2–2·2 g. (74–81·5%).

### Reviews

(a) "Meso-Ionic Compounds", W. Baker and W. D. Ollis, *Quart. Rev.*, 1957, **11**, 15.
(b) "Une Famille originale; les sydnones", Y. Noel, *Bull. Soc. chim. France*, 1964, 173.
(c) "The Chemistry of the Sydnones", F. H. C. Stewart, *Chem. Rev.*, 1964, **64**, 129.

### References

1. W. Baker and W. D. Ollis, *Chem. and Ind.*, 1955, 910.
2. J. C. Earl and A. W. Mackney, *J. Chem. Soc.*, 1935, 899; R. A. Eade and J. C. Earl, *J. Chem. Soc.*, 1946, 591.
3. W. Baker, W. D. Ollis, and V. D. Poole, *J. Chem. Soc.*, 1949, 307.

## 3,4-DIMETHYL-1,2,5-OXADIAZOLE (3,4-DIMETHYLFURAZAN)

1,2,5-Oxadiazoles are usually prepared by dehydration of α-diketone dioximes under basic conditions.[1] Succinic anhydride has, however, been found to be an effective dehydrating agent.[2]

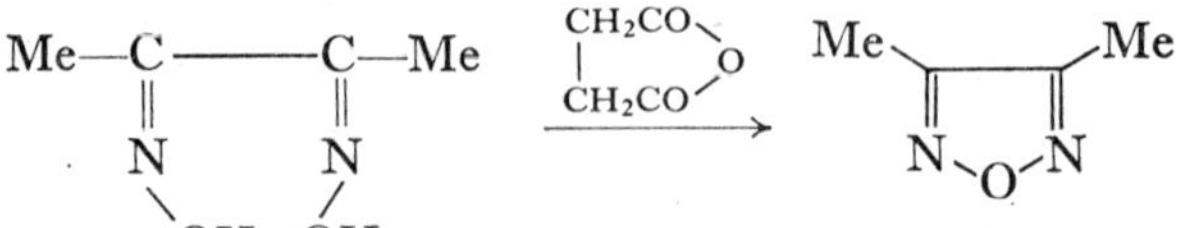


I

### 3,4-Dimethyl-1,2,5-oxadiazole (I)

Into a 100-ml. round-bottomed flask equipped with a reflux condenser, place an intimately ground mixture of dimethylglyoxime (11·6 g.) and succinic anhydride (10 g.). Heat the flask in an oil-bath or Woods-metal bath until an exothermic reaction commences (at ca. 160°), then remove the flask from the heating-bath and allow the reaction to subside (ca. 15 min.). Distil the resulting mixture and collect the fraction b.p. 150–200°. Dissolve the distillate in ether, dry ($MgSO_4$), and evaporate off the solvent. Transfer the residual oil into a 25-ml. pear-shaped flask fitted with a short fractionation column, and distil at atmospheric pressure, neglecting the fore-run (ca. 3 ml.), b.p. 60–110°.

3,4-Dimethyl-1,2,5-oxadiazole is obtained as a colourless oil, b.p. 152–158°.

Yield = 6·5–7·5 g. (66·5–76·5%).

### Spectra

(a) Ultraviolet.[2]
(b) Infrared.[3]
(c) Proton magnetic resonance.[4]

### Reviews

(a) "Oxadiazoles and Related Compounds", "The Chemistry of Heterocyclic Compounds", ed. A. Weissberger, Vol. 17, p. 235.
(b) "Oxadiazoles", "Heterocyclic Compounds", ed. R. C. Elderfield, Vol. 7, p. 464.
(c) "1,2,5-Oxadiazoles (furazans)", "Chemistry of Carbon Compounds", ed. E. H. Rodd, Vol. IV A, p. 466.

### References

1. L. C. Behr and J. T. Brent, *Org. Synth.*, 1963, Coll. Vol. IV, 342.
2. H. E. Acly and H. S. French, *J. Amer. Chem. Soc.*, 1927, **49**, 847.
3. M. Milone and E. Borello, *Gazzetta*, 1951, **81**, 368, 677.
4. G. Englert, *Z. analyt. Chem.*, 1961, **181**, 447.

## 2,5-DIPHENYL-1,3,4-OXADIAZOLE

1,3,4-Oxadiazoles are generally obtained from the cyclization of *N,N'*-diacylhydrazines.[1] However, polyphosphoric acid (PPA) has been found to cyclize quantitatively monoacyl hydrazines to the corresponding 1,3,4-oxadiazoles.[2] For example, 2,5-diphenyl-1,3,4-oxadiazole (II) is obtained from benzohydrazide (I).

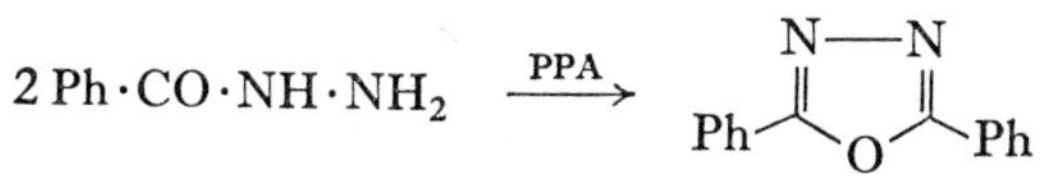


I II

### Benzohydrazide (I)

To a solution of hydrazine hydrate (10 ml.) in ethanol (20 ml.) contained in a 100-ml. round-bottomed flask, add ethyl benzoate (10 g.; 9·5 ml.) and heat the mixture under reflux for 1 hr. Cool, filter the solution and wash the product with ice-cold ethanol (2 × 25 ml.).

Benzohydrazide is obtained as a white crystalline solid, m.p. 112°, sufficiently pure for direct use in the next stage.

Yield = 7–8 g. (77–87·5%).

### 2,5-Diphenyl-1,3,4-oxadiazole (II)

To benzohydrazide (2 g.), contained in the apparatus shown, add polyphosphoric acid (40 g.) and stir the mixture at 170° for 40 min., using phenetole,

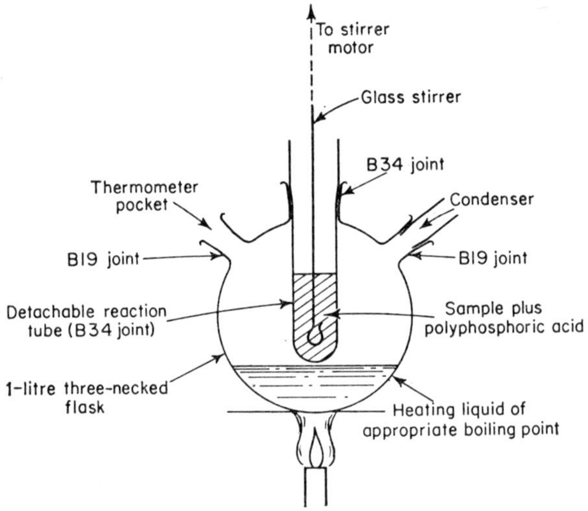


b.p. 170°, as heating-bath liquid. Cool the solution to ca. 80°, pour the viscous mass into ice–water (200 ml.), and stir until a white solid precipitates. Filter off the product, wash with cold water and dry at 100°.

2,5-Diphenyl-1,3,4-oxadiazole crystallizes from ethanol (ca. 10 ml./g.) as colourless plates, m.p. 137–138°.

Yield = 1·4–1·5 g. (86–92%).

**Spectra**

(a) Ultraviolet.[3]
(b) Infrared.[4]

## Reviews

(a) "Oxadiazoles and Related Compounds", "The Chemistry of Heterocyclic Compounds", ed. A. Weissberger, Vol. 17, p. 235.
(b) "Oxadiazoles", "Heterocyclic Compounds", ed. R. C. Elderfield, Vol. 7, p. 462.
(c) "1,3,4-Oxadiazoles", "Chemistry of Carbon Compounds", ed. E. H. Rodd, Vol. IV A, p. 471.
(d) "The Chemistry of 1,3,4-Oxadiazole Derivatives", E. P. Nesynov and A. P. Grekov, *Russ. Chem. Rev.*, 1964, **33**, 508.
(e) "Polyphosphoric Acid", see review references on p. 78.

## References

1. R. Stollé, *J. prakt. Chem.*, 1903, **68**, 130; 1904, **69**, 503, 506.
2. C. T. Elston, Doctoral Dissertation, Univ. of Illinois, 1954.
3. D. G. Ott, V. N. Kerr, F. N. Hayes, and E. Hansbury, *J. Org. Chem.*, 1960, **25**, 872.
4. E. Müller and D. Ludsteck, *Chem. Ber.*, 1955, **88**, 921.

## 2,4-DIMETHYLTHIAZOLE

With phosphorus pentasulphide, acetamide yields thioacetamide, and this condenses with chloroacetone to yield 2,4-dimethylthiazole (I).[1,2]

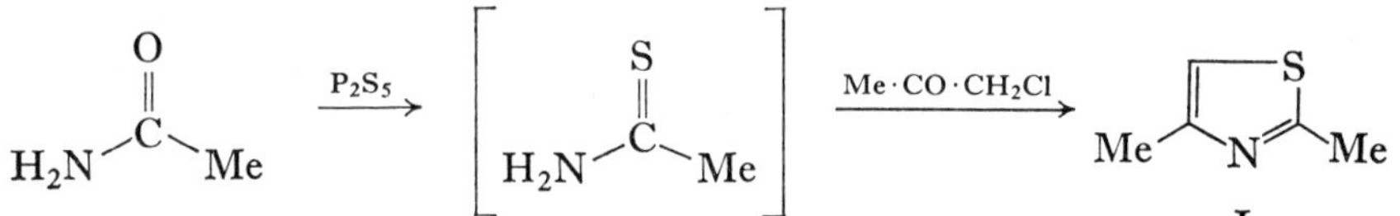


I

### 2,4-Dimethylthiazole (I)

In a 250-ml., 2-necked round-bottomed flask fitted with reflux condenser and dropping-funnel, place dry benzene (20 ml.), and add a finely ground, rapidly prepared mixture of acetamide (30 g.) and phosphorus pentasulphide (20 g.). In the dropping funnel, place a solution of chloroacetone (46·5 g.; 40 ml.) in benzene (15 ml.), and add ca. 5 ml. of this solution into the flask. Cautiously heat the mixture until an exothermic reaction occurs, then remove the source of heat. When the reaction has subsided, add a second 5 ml. portion of the solution from the dropping funnel and continue heating carefully removing the source of heat if the reaction becomes too vigorous. Repeat the procedure until all the chloroacetone solution has been added (ca. 20 min.), then heat under reflux for 0·5 hr. At the end of this period, cool and add water (75 ml.). After 15 min., transfer into a separating-funnel, and run off the *required* aqueous layer, and wash the benzene layer with water (25 ml.). Combine the aqueous extracts and discard the benzene solution. Basify the aqueous solution with 4N sodium hydroxide, and extract the liberated product with ether (3 × 60 ml.). Combine the extracts, dry ($MgSO_4$), and after evaporating off the ether, distil the residual oil at atmospheric pressure.

2,4-Dimethylthiazole is obtained as a foul-smelling, pale-yellow liquid, b.p. 141–144°.

Yield = 19–21 g. (33–37%).

### Spectra

(a) Infrared.[3]
(b) Proton magnetic resonance.[4]

### Reviews

(a) "Thiazoles", "Heterocyclic Compounds", ed. R. C. Elderfield, Vol. 5, p. 496.
(b) "Thiazoles", "Chemistry of Carbon Compounds", ed. E. H. Rodd, Vol. IV A, p. 385.
(c) "The Preparation of Thiazoles", R. H. Wiley, D. C. England, and L. C. Behr, *Org. Reactions*, 1951, **6**, 367.

### References

1. E. Merck, German Patent 670,131, (*Chem. Abs.* 1939, **33**, 2909).
2. G. Schwarz, *Org. Synth.*, 1955, Coll. Vol. III, p. 332.
3. P. Bassignana, C. Cogrossi, and M. Gandino, *Spectrochim. Acta*, 1963, **19**, 1885.
4. A. Taurins and W. G. Schneider, *Canad. J. Chem.*, 1960, **38**, 1237.

## 2-METHYL-Δ2-THIAZOLINE

β-Hydroxyalkylamides are cyclized in the presence of phosphorus pentasulphide to Δ2-thiazolines. *N*-Acetylethanolamine (I) (p. 32) is thus converted into 2-methyl-Δ2-thiazoline (II).[1]

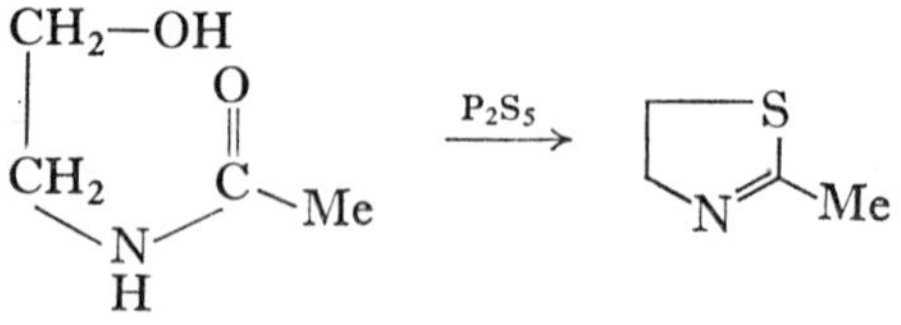


I II

### 2-Methyl-Δ2-thiazoline (II)

Into a **500-ml.** round-bottomed flask, place phosphorous pentasulphide (13.3 g.). Using the continuous distillation procedure,[2] cautiously add *N*-acetylethanol-amine (30.9 g.) from the dropping funnel until a vigorous exothermic reaction occurs, and as a result of which a certain amount of the product distils. When the reaction moderates, add any remaining *N*-acetylethanolamine, then replace the dropping-funnel by a thermometer and heat the flask gently to complete the distillation, collecting the product b.p. 145–170°.

To the distillate, add potassium hydroxide pellets (10 g.), and transfer the resulting 2-phase liquid to a separating funnel. Separate the 2 phases, and extract the caustic phase with ether (15 ml.). Add this extract to the organic phase, dry ($MgSO_4$), and after evaporation of the ether, distil the residual oil at atmospheric pressure.

2-Methyl-$\Delta^2$-thiazoline is obtained as an almost colourless, foul-smelling liquid, b.p. 145–148°.

Yield = 10–12 g. (33–40%).

## Spectra

(a) Ultraviolet.[3]
(b) Infrared.[4]
(c) Proton magnetic resonance.[5]

## Reviews

(a) "Thiazolines", "Heterocyclic Compounds", ed. R. C. Elderfield, Vol. 5, p. 677.
(b) "Thiazolines", "Chemistry of Carbon Compounds", ed. E. H. Rodd, Vol. IV A, p. 403.

## References

1. H. Wenker, *J. Amer. Chem. Soc.*, 1935, **57**, 1079.
2. A. I. Vogel, "A Text-book of Practical Organic Chemistry", Longmans, Green & Co., London, 1956, 3rd edn., p. 88.
3. R. Kuhn and F. Drawert, *Annalen*, 1954, **590**, 55.
4. P. Bassignana, C. Cogrossi, and M. Gandino, *Spectrochim. Acta*, 1963, **19**, 1885.
5. M. A. Weinberger and R. Greenhalgh, *Canad. J. Chem.*, 1963, **41**, 1038.

# 4 BENZODERIVATIVES OF FIVE-MEMBERED SYSTEMS CONTAINING MORE THAN ONE HETEROATOM

## BENZIMIDAZOLE

Benzimidazoles are formed by the action of an acylating agent on *o*-phenylenediamine.[1,2] Benzimidazole is obtained by heating under reflux a solution of *o*-phenylenediamine in formic acid. The reaction proceeds via the acyl derivative, which cyclizes under the influence of excess of acid.

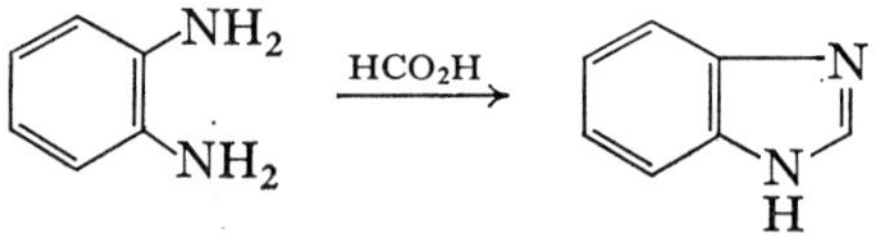


I

### Benzimidazole (I)

In a 50-ml. round-bottomed flask, heat a solution of *o*-phenylenediamine (2·7 g.) in 98% formic acid (10 ml.), under reflux for 1 hr. Cool the reaction mixture, dilute with cold water (10 ml.), and basify the solution by addition of solid sodium carbonate. Filter off the crude product, and wash with ice–water (100 ml.).

Benzimidazole crystallizes from water (ca. 12 ml./g.) as white prisms, m.p. 170–171°.

Yield = 2·4–2·6 g. (81·5–88%).

### Spectra

(a) Ultraviolet.[3]
(b) Infrared.[4]
(c) Proton magnetic resonance.[5]

## Reviews

(a) "The Chemistry of the Benzimidazoles", J. B. Wright, *Chem. Rev.*, 1951, **48**, 397.
(b) "Advances in the Chemistry of Imidazoles", A. F. Pozharskii, A. D. Garnovskii, and A. M. Simonov, *Russ. Chem. Rev.*, 1966, **35**, 122.
(c) "Imidazoles and Condensed Imidazoles", "Heterocyclic Compounds", ed. R. C. Elderfield, Vol. 5, p. 194.
(d) "Benziminazoles", "Chemistry of Carbon Compounds", ed. E. H. Rodd, Vol IV A, p. 321.

## References

1. A. Ladenburg, *Ber.*, 1875, **8**, 677.
2. M. A. Phillips, *J. Chem. Soc.*, 1928, 172, 2393; 1929, 2820.
3. E. C. Fisher and M. M. Joulliè, *J. Org. Chem.*, 1958, **23**, 1944.
4. K. J. Morgan, *J. Chem. Soc.*, 1961, 2343.
5. P. J. Black and M. L. Heffernan, *Austral. J. Chem.*, 1962, **15**, 862.

## 1,2-DIHYDRO-2-OXOBENZIMIDAZOLE (BENZIMIDAZOLONE)

On heating in an inert solvent, acid azides undergo a Curtius re-arrangement to the corresponding isocyanate, which can react with a primary amine to yield an *N,N'*-disubstituted urea. 1,2-Dihydro-2-oxobenzimidazole (III), which may be considered to be a cyclic *N,N'*-disubstituted urea, is easily obtained by the re-arrangement of anthranilic acid azide (I)[1] (see p. 119). The initially formed isocyanate (II) undergoes an internal addition with the adjacent amine group.

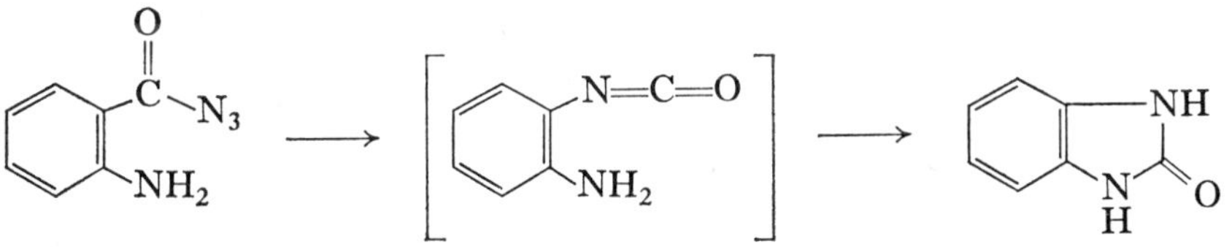


I II III

### 1,2-Dihydro-2-oxobenzimidazole (III)

In a 100-ml. round-bottomed flask, heat a mixture of anthranilic acid azide (1 g.) in dry toluene (50 ml.), under reflux for 3 hr. Cool the reaction mixture, and filter off the precipitate.

1,2-Dihydro-2-oxobenzimidazole crystallizes from ethanol (ca. 20 ml./g.) as colourless prisms, m.p. 305–306°.

Yield = 0·7–0·75 g. (84·5–90·5%).

For a discussion on the tautomerism of benzimidazolones see ref. 2.

**Spectra**

(a) Ultraviolet.[3]
(b) Infrared.[3,4]

**Reviews**

"Benzimidazolones", see review references on p. 43.

**References**


1. G. Heller and A. Siller, *J. prakt. Chem.*, 1927, **116**, 9.
2. A. R. Katritzky and J. M. Lagowski, *Adv. Heterocyclic Chem.*, 1963, **2**, 50.
3. D. Harrison and A. C. B. Smith, *J. Chem. Soc.*, 1959, 3157.
4. S. F. Mason, *J. Chem. Soc.*, 1957, 4874.


## 6-NITROINDAZOLE

Indazoles may be prepared, along with hydroxytoluenes, by the diazotization and subsequent decomposition of an *o*-toluidine.[1] The diazonium group activates the adjacent methyl group towards electrophilic attack, the effect being enhanced by the presence of an electron-attracting group in the *para*-position. In this example, 6-nitroindazole (II) is obtained by decomposition of diazotized 2-amino-4-nitrotoluene (I).[2]

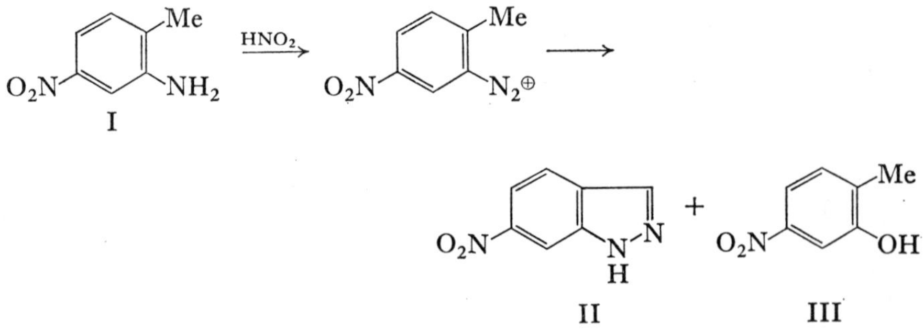


### 6-Nitroindazole (II)

Dissolve 2-amino-4-nitrotoluene (10 g.) in concentrated sulphuric acid (20 ml.), then cautiously pour the acid solution into ice–water (300 ml.), and stir vigorously, so that the precipitate of amine sulphate is finely dispersed. To the suspension, add a solution of sodium nitrite (4·7 g.) in water (15 ml.) and stir until all the amine sulphate dissolves. Filter to remove any insoluble impurities, and warm the filtrate in a 500-ml. conical flask on a water-bath. When smooth evolution of nitrogen is noted, heat for a further 3 hr. on a water-bath at 100°.

Cool the solution and filter off the bright-yellow precipitate of 6-nitroindazole (II) and 2-methyl-5-nitrophenol (III). Dissolve the yellow residue in hot water (300 ml.) and 30% sodium hydroxide (10 ml.), and then rapidly filter the hot solution. Cool the filtrate and precipitate the product by adding an excess of solid carbon dioxide. Filter off the product and dry at 100°.

6-Nitroindazole crystallizes from petroleum ether (b.p. 100–120°; ca. 6 ml./g.) as bright-yellow needles, m.p. 178–180°.

Yield=4·8–5·5 g. (45–51%).

## Spectra

(a) Ultraviolet.[3]
(b) Infrared.[4]
(c) Proton magnetic resonance.[5]

## Reviews

(a) "Indazoles", "Heterocyclic Compounds", ed. R. C. Elderfield, Vol. 5, p. 162.
(b) "Indazoles or Benzopyrazoles", "Chemistry of Carbon Compounds", ed. E. H. Rodd, Vol. IV A, p. 278.

## References

1. E. Bamberger and A. von Goldberger, *Annalen*, 1899, **305**, 339.
2. O. N. Witt, E. Noelting, and E. Grandmougin, *Ber.*, 1890, **23**, 3635.
3. A. Bellotti and G. Pappalardo, *Boll. Sci. Fac. Chim. ind. Bologna*, 1958, **16**, 29.
4. D. G. O'Sullivan, *J. Chem. Soc.*, 1960, 3278.
5. P. J. Black and M. L. Heffernan, *Austral. J. Chem.*, 1963, **16**, 1051.

# 1-AMINOBENZOTRIAZOLE

Benzotriazoles result from the cyclization of *o*-aminobenzenediazonium salts, (see also p. 47). Thus the cyclization of the diazonium salt (II) leads to diethyl mesoxylidene 1-aminobenzotriazole (III), which on hydrolysis yields 1-aminobenzotriazole (IV).[1] Diazotization of *o*-nitroaniline followed by interaction of the diazonium salt with diethyl malonate leads to a Japp–Klingemann reaction[2] and formation of diethyl mesoxalate *o*-nitrophenylhydrazone (I). Catalytic reduction of the nitro group and subsequent diazotization of the resulting amine yields the diazonium salt (II), which cyclizes spontaneously.

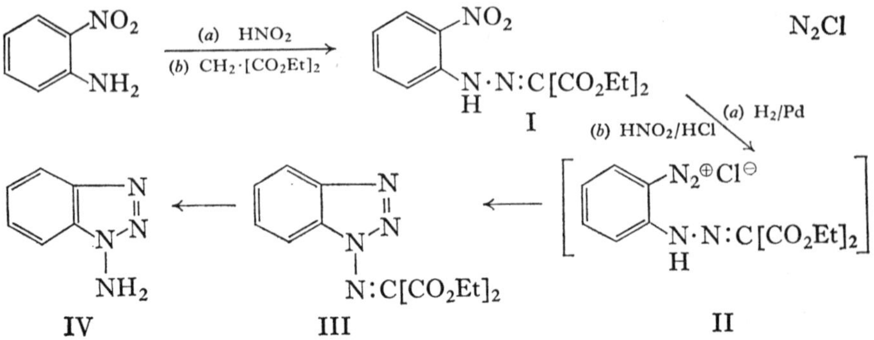


## Diethyl mesoxalate *o*-nitrophenylhydrazone (I)

Grind *o*-nitroaniline (20·7 g.) with concentrated hydrochloric acid (45 ml.) and transfer the resulting fine suspension of *o*-nitroaniline hydrochloride into a 250-ml. beaker, and add water (100 ml.). Magnetically stir the suspension and with ice–salt cooling, add dropwise a solution of sodium nitrite (12·1 g.) in water (100 ml.), maintaining the temperature of the reaction between 0 and 5°. Transfer the resulting clear solution into a 250-ml. dropping-funnel.

In a 500-ml. beaker, mechanically stir an emulsion of diethyl malonate (24 g.; 23 ml.) in water (100 ml.) and with ice–salt cooling, add dropwise the *o*-nitrobenzenediazonium chloride solution over 1 hr. at $5 \pm 2°$, keeping the reaction alkaline to litmus by portionwise addition of sodium carbonate. When addition is complete, filter the mixture, wash the residual product with water ($3 \times 25$ ml.) and dry as far as possible by suction.

Diethyl mesoxalate *o*-nitrophenylhydrazone crystallizes from ethanol (ca. 1 ml./g.) as yellow needles, m.p. 72–73°.

Yield=35–37·5 g. (75–81%).

## Diethyl mesoxylidene 1-aminobenzotriazole (III)

To a suspension of diethyl mesoxalate *o*-nitrophenylhydrazone (15·4 g.) in methanol (160 ml.) contained in a 250-ml. hydrogenation flask, add 10% palladium–charcoal catalyst (1 g.). Fit the flask to an atmospheric hydrogenation apparatus and allow to shake in a hydrogen atmosphere until the theoretical quantity (3400 ml.) of hydrogen has been taken up (ca. 2 hr.). Some cooling may be necessary during the reduction.

Remove the flask from the hydrogenation apparatus, filter off the catalyst and to the filtrate add a solution of sodium nitrite (4 g.) in water (25 ml.). Add the resulting solution dropwise over 20 min. to a mechanically stirred mixture of concentrated hydrochloric acid (20 ml.) and water (25 ml.) maintained at 0–5°. Filter off the brown precipitate and dry by suction.

Diethyl mesoxylidene 1-aminobenzotriazole crystallizes from ethanol (ca. 2·5 ml./g.) (decolourizing carbon), as large colourless crystals, m.p. 98–99°.

Yield=10·5–11·5 g. (72·5–80%).

## 1-Aminobenzotriazole (IV)

In a 250-ml. conical flask, magnetically stir diethyl mesoxylidene 1-amino-benzotriazole (10 g.) in concentrated hydrochloric acid (50 ml.) until the solid has dissolved (ca. 3 hr.). To the resulting solution (which may contain a small amount of undissolved solid), add water (50 ml.), then wash with ether (3 × 25 ml.) and discard the ethereal washings. Carefully basify the acidic solution with a saturated solution of potassium carbonate, and repeatedly extract the resulting solution with ether (ca. 8 × 40 ml.). Combine the extracts, dry ($MgSO_4$) and evaporate off the ether to obtain the crude product as a colourless oil that readily solidifies.

1-Aminobenzotriazole crystallizes from a 60–40 mixture of benzene–petroleum ether (b.p. 60–80°) (ca. 20 ml./g.) as white needles, m.p. 81–82°.

Yield=3·45–3·9 g. (75–85%).

### Reviews

(a) "Benzotriazoles", see review references on p. 48.
(b) "Japp-Klingemann reaction", R. R. Phillips, *Org. Reactions*, 1959, **10**, 143.

### References

1. C. D. Campbell and C. W. Rees, University of Leicester, private communication.
2. F. R. Japp and F. Klingemann, *Ber.*, 1888, **21**, 549.

## BENZOTRIAZOLE

Diazotization of *o*-phenylenediamine in dilute acid solution yields benzotriazole.[1] The reaction proceeds via the mono-diazonium salt (I), which then couples with the adjacent amino-group to give the internal diazoamino compound, benzotriazole (II).

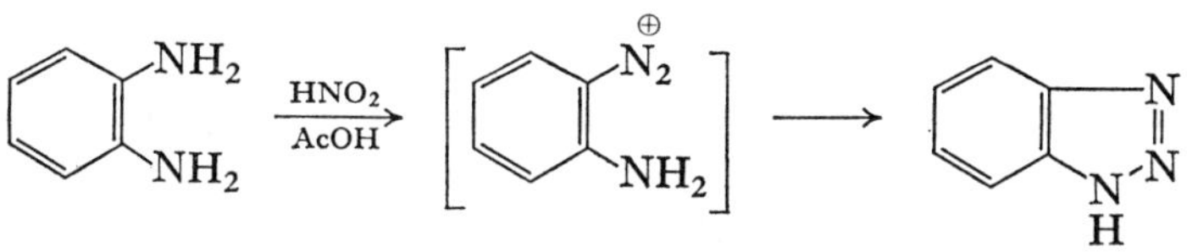


I II

### Benzotriazole (II)

To a solution of *o*-phenylenediamine (5·4 g.) in a mixture of glacial acetic acid (12 ml.) and water (25 ml.) contained in a 250-ml. beaker, add a solution of sodium nitrite (3·5 g.) in water (10 ml.). Cool the resulting solution, which will have attained a temperature of ca. 70°, in an ice-bath and neutralize to pH 7 by adding 2N ammonium hydroxide. Stir the mixture until a solid product is obtained, then filter, and wash the brown residue with ice–water (100 ml.). Dry the product at 80°.

Benzotriazole crystallizes from benzene (ca. 5 ml./g.), (decolourizing carbon) as colourless needles, m.p. 100–101°.

Yield = 4–4·6 g. (67–77%).

### Spectra

(a) Ultraviolet.[2]
(b) Infrared.[3,4]
(c) Proton magnetic resonance.[5]

### Reviews

(a) "Monocyclic Triazoles and Benzotriazoles", "Heterocyclic Compounds", ed. R. C. Elderfield, Vol. 7, p. 384.
(b) "Benzotriazoles", "Chemistry of Carbon Compounds", ed. E. H. Rodd, Vol. IV A, p. 449.

### References

1. A. Ladenburg, *Ber.*, 1876, **9**, 219.
2. B. W. Ashton and H. Suschitzky, *J. Chem. Soc.*, 1957, 4559.
3. K. J. Morgan, *J. Chem. Soc.*, 1961, 2343.
4. D. G. O'Sullivan, *Spectrochim. Acta*, 1960, **16**, 762.
5. P. J. Black and M. L. Heffernan, *Austral. J. Chem.*, 1962, **15**, 862.

## 1-(*p*-NITROPHENYL)-4,5,6,7-TETRAHYDROBENZOTRIAZOLE

The interaction of 1-morpholinocyclohexene,[1] (the morpholine enamine of cyclohexanone) (I) and *p*-nitrophenylazide[2] (II) leads to 1-(*p*-nitrophenyl)-3a,4,5,6,7,7a-hexahydro-7a-morpholinobenzotriazole (III), which on hydrolysis yields 1-(*p*-nitrophenyl)-4,5,6,7-tetrahydrobenzotriazole (IV).[3]

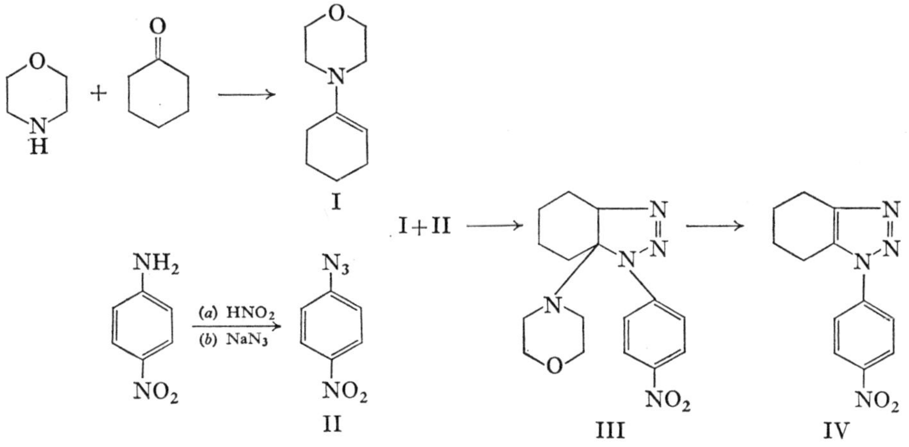

## 1-Morpholinocyclohexene (I)

In a 100-ml. round-bottomed flask, dissolve cyclohexanone (9·8 g.; 10·3 ml.), morpholine (13 g.; 13 ml.) and *p*-toluenesulphonic acid (0·2 g.) in benzene (40 ml.). Heat the solution using a Dean and Stark apparatus[4] and reflux condenser for 3 hr. when water (ca. 1·5 ml.) separates.

Carefully evaporate off the benzene and excess of morpholine, then distil the residue with water-pump vacuum.

1-Morpholinocyclohexene is obtained as a colourless liquid, b.p. 120–122°/12 mm.

Yield = 12–15 g. (72–90%).

## *p*-Nitrophenylazide (II)

In a 250-ml. conical flask, dissolve *p*-nitroaniline (5 g.) in a mixture of concentrated hydrochloric acid (20 ml.) and water (45 ml.), and cool the solution to 0°. With ice–salt cooling, rapidly add a cooled solution of sodium nitrite (2·5 g.) in water (5 ml.), keeping the internal temperature between 0 and 5° during addition. Allow the resulting solution to stand for 10 min. at 0°, then stir this slowly into a solution of sodium acetate (14 g.) and sodium azide (2·2 g.) in ice–water (60 g.) contained in a 500-ml. conical flask.

**Note**

All operations involving sodium azide must be carried out in an efficient fume cupboard, since aqueous solutions of this compound evolve hydrazoic acid, which is highly toxic.

Filter off the precipitated product, wash with water, and dry in a *vacuum desiccator.*

*p*-Nitrophenylazide crystallizes from ethanol (ca. 3 ml./g.) as straw-coloured plates, m.p. 72–73°. Dry in a *vacuum desiccator.*

Yield=4–5 g. (68–85%).

### Note

Nearly all azides decompose violently when heated alone either in the solid or the liquid state. The prepared azide must *not*, therefore, be dried over a flame or in an oven.

## 1-(*p*-Nitrophenyl)-3a,4,5,6,7,7a-hexahydro-7a-morpholino-benzotriazole (III)

To a solution of *p*-nitrophenylazide (3·28 g.) in dry chloroform (50 ml.) in a 100-ml. round-bottomed flask, add 1-morpholinocyclohexene (3·34 g.) and heat the mixture under reflux for 3 hr. Evaporate off the chloroform to obtain the crude product as a fawn residue.

1-(*p*-Nitrophenyl)-3a,4,5,6,7,7a-hexahydro-7a-morpholinobenzotriazole crystallizes from benzene (ca. 20 ml./g.) as clumps of pale-yellow prisms, m.p. 173–174° (decomp.).

Yield=4–5 g. (60–76%).

## 1-(*p*-Nitrophenyl)-4,5,6,7-tetrahydrobenzotriazole (IV)

In a 100-ml. round-bottomed flask, heat 1-(*p*-nitrophenyl)-3a,4,5,6,7,7a-hexahydro-7a-morpholinobenzotriazole (2 g.) with a mixture of ethanol (10 ml.) and 2N hydrochloric acid (15 ml.) for 0·5 hr., then evaporate the solution to dryness. To the residue, add 2N sodium hydroxide (15 ml.) and filter off the precipitated product.

1-(*p*-Nitrophenyl)-4,5,6,7-tetrahydrobenzotriazole crystallizes from ethanol (ca. 50 ml./g.) (decolourizing carbon) as long yellow needles, m.p. 224–225°.

Yield=0·5–0·7 g. (34–47%).

### Reviews

(a) "Monocyclic Triazoles and Benzotriazoles", "Heterocyclic Compounds", ed. R. C. Elderfield, Vol. 7, p. 384.
(b) "Triazoles", "Chemistry of Carbon Compounds", ed. E. H. Rodd, Vol. IV A, p. 439.
(c) "Benzotriazoles", "Chemistry of Carbon Compounds", ed. E. H. Rodd, Vol. IV A, p. 449.
(d) "Enamines", see review reference on p. 97.
(e) "Azides", see review reference on p. 9.

### References

1. S. Hünig, E. Benzing, and E. Lucke, *Chem. Ber.*, 1957, **90**, 2833.
2. E. Noelting and K. Kohn, *Chem. Zentr.*, 1894, **18**, 1095.

3. R. Fusco, G. Bianchetti, and D. Pocar, *Gazzetta*, 1961, **91**, 849, (*Chem. Abs.*, 1962, **56**, 14019).
4. A. I. Vogel, "A Text-book of Practical Organic Chemistry", Longmans, Green & Co., London, 3rd edn., 1956, p. 429.

## 3-METHYL-5-NITROBENZISOXAZOLE

Careful nitration of *o*-bromoacetophenone[1] gives 2-bromo-5-nitroacetophenone (III).[2] Treatment of the oxime of (III) with alkali brings about an intramolecular nucleophilic displacement of the activated bromine atom by the oxime anion. The resulting benzisoxazole is more easily obtained from the β-oxime (V), in which the nucleophile is closer to the substitution centre, than it is in the α-oxime (IV).

The α- (IV) and β- (V) forms of the oxime can be separated,[1,3] if required, by fractional crystallization.

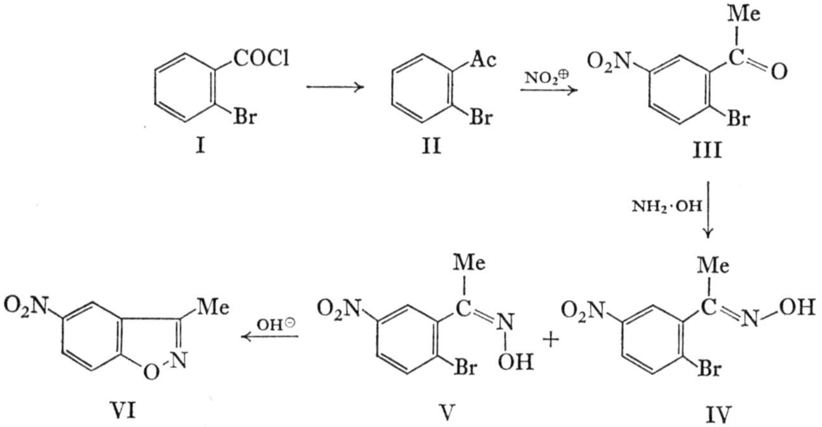


### *o*-Bromoacetophenone (II)

*o*-Bromoacetophenone is prepared by the following route.[1]

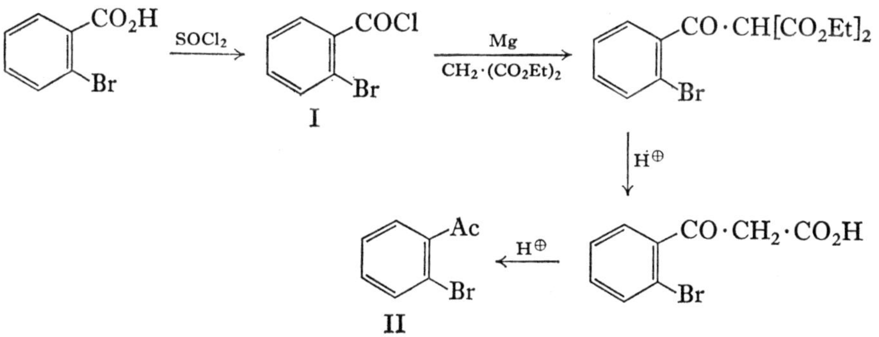

## *o*-Bromobenzoyl chloride (I)

To thionyl chloride (75 ml.) contained in a 250-ml. 2-necked, round-bottomed flask, fitted with a reflux condenser, add *o*-bromobenzoic acid (20 g.). Heat the mixture under reflux for 3 hr., then distil off the excess of thionyl chloride (b.p. 78–79°) from a water-bath at 100°. Transfer the residual oil into a 25-ml. pear-shaped flask and distil under water-pump vacuum.

*o*-Bromobenzoyl chloride is obtained as a colourless oil, b.p. 132–134°/20 mm. Yield = 17–17·5 g. (77·5–80%).

**Note**

The product should be kept in a stoppered flask in order to prevent hydrolysis.

## *o*-Bromoacetophenone (II)

In a dry 500-ml. 3-necked, round-bottomed flask, fitted with a 50-ml. dropping-funnel, reflux condenser and stirrer, place clean magnesium turnings (3·15 g.), ethanol (5 ml.), and carbon tetrachloride (0·5 ml.). When the vigorous reaction commences (1–2 min.), add chlorobenzene (20 ml.), and then diethyl malonate (20·5 g.; 20·4 ml.) at a rate such that the temperature of the reaction mixture remains below 70°. After ester addition is complete, heat the mixture on a water-bath at 70–75° for 3 hr. Cool the solution and add slowly a solution of *o*-bromobenzoyl chloride (16·7 g.) in chlorobenzene (35 ml.), at a rate such that the temperature of the reaction mixture remains below 35°. Allow the mixture to stir overnight at room temperature.

To the resulting yellow viscous mass, add 25% sulphuric acid (40 ml.), and heat on a water-bath at 100° for 1 hr. Cool the solution, transfer into a separating funnel, and run the lower organic layer into a 100-ml. round-bottomed flask. Remove the excess of chlorobenzene (ca. 50 ml.) under reduced pressure, and dissolve the residual pale-yellow oil in a mixture of glacial acetic acid (40 ml.) and 20% sulphuric acid (40 ml.). Heat the solution overnight under reflux.

Dilute the reaction mixture with water (50 ml.), transfer into a separating funnel, and extract the solution with ether (2 × 50 ml.). Wash the combined ethereal extracts with cold water (2 × 50 ml.), dry ($MgSO_4$) and evaporate off the ether to obtain the crude product as a yellow oil. Distil under water-pump vacuum.

*o*-Bromoacetophenone is obtained as a colourless oil, b.p. 117–118°/12 mm. Yield = 12–12·5 g. (79–82·5%).

## 2-Bromo-5-nitroacetophenone (III)

Cool fuming nitric acid (sp. gr. 1·52; 25 ml.) in a 250-ml. conical flask to −10° in a solid carbon dioxide–acetone bath. Magnetically stir the acid solution and add *o*-bromoacetophenone (6·1 g.) dropwise, at a rate such that the temperature of the reaction mixture remains between −5 and −10°. When the addition is complete, continue stirring at −5° for 1 hr., then cautiously pour the yellow reaction mixture into cold water (150 ml.). Filter off the product and wash with cold water (4 × 60 ml.).

2-Bromo-5-nitroacetophenone crystallizes from ethanol (ca. 5 ml./g.) as pale-yellow needles, m.p. 86–88°.

Yield = 5·6–6 g. (75–80%).

### 3-Methyl-5-nitrobenzisoxazole (VI)

Dissolve 2-bromo-5-nitroacetophenone (2·45 g.) in ethanol (20 ml.) and add a solution of hydroxylamine hydrochloride (0·75 g.) in water (6 ml.). Heat the resulting mixture on a water-bath at 100° for 0·5 hr., then add 0·5N ethanolic potassium hydroxide (20 ml.) and heat under reflux for a further 1 hr. Allow the deep-red solution to cool and filter off the product.

3-Methyl-5-nitrobenzisoxazole crystallizes from ethanol (ca. 20 ml./g.) as pale-orange needles, m.p. 127–128°.

Yield = 0·5–0·7 g. (28–39·5%).

### Reviews

(a) "Isoxazoles and Related Compounds", "The Chemistry of Heterocyclic Compounds", ed. A. Weissberger, Vol. 17, p. 159.
(b) "Isoxazoles", "Heterocyclic Compounds", ed. R. C. Elderfield, Vol. 5, p. 452.
(c) "Benzisoxazoles (Indoxazenes)", "Chemistry of Carbon Compounds", ed. E. H. Rodd, Vol. IV A, p. 349.

### References

1. J. Klein and E. D. Bergmann, *J. Org. Chem.*, 1957, **22**, 1019.
2. J. Meisenheimer, P. Zimmerman, and U. von Kummer, *Annalen*, 1926, **446**, 216.
3. N. H. P. Smith, *J. Chem. Soc.*, 1961, 4209.

## 3-PHENYLBENZO-2,1-ISOXAZOLE (3-PHENYLANTHRANIL)

Decomposition[1] of *o*-azidobenzophenone (I) yields 3-phenylbenzo-2,1-isoxazole (II). This reaction, which is similar to the decomposition of *o*-nitrophenyl azide (see p. 57), probably involves an intramolecular displacement of nitrogen from the azido group by the adjacent carbonyl group, rather than a nitrene intermediate.

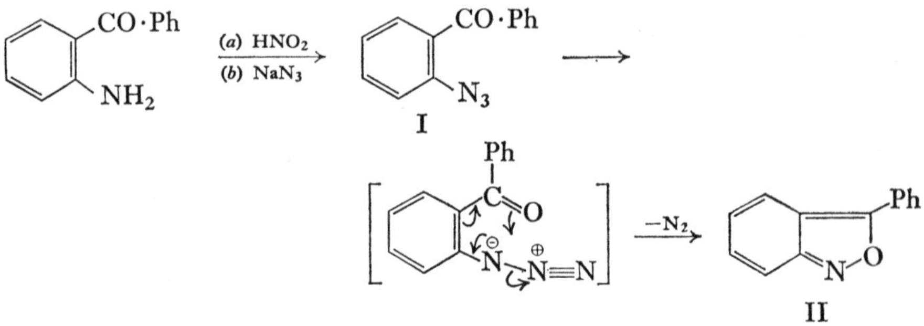


## *o*-Azidobenzophenone (I)

Cool a solution of *o*-aminobenzophenone[2] (4 g.) in 6N hydrochloric acid (40 ml.) in an ice-bath, and add with stirring, a solution of sodium nitrite (2 g.) in water (10 ml.). Allow to stir for 10 min., then add the mixture slowly to a solution of sodium azide (2·5 g.) and sodium acetate (60 g.) in cold water (150 ml.), contained in a 500-ml. conical flask.

**Note**

All operations involving sodium azide must be carried out in an efficient fume cupboard, since aqueous solutions of this compound evolve hydrazoic acid ($HN_3$), which is highly toxic.

Stir the solution for 10 min., then transfer into a separating funnel, and extract the oily product with ether (2 × 50 ml.). Wash the combined ethereal extracts with cold water (3 × 50 ml.), dry ($MgSO_4$), and evaporate off the ether on a water-bath at 50–70°. (Higher temperatures must be avoided as the azide tends to decompose; see warning on p. 8.)

*o*-Azidobenzophenone is obtained as a pale-yellow oil, sufficiently pure for direct use in the next stage.

Yield = 3·8–4·3 g. (85–96·5%).

## 3-Phenylbenzo-2,1-isoxazole (II)

To refluxing chlorobenzene (160 ml.) contained in a 500-ml., 2-necked round-bottomed flask fitted with dropping-funnel and reflux condenser, add dropwise over 10 min., a solution of *o*-azidobenzophenone (3·5 g.) in chlorobenzene (40 ml.). Heat the mixture under reflux for 1 hr., and then distil off the excess of solvent (ca. 190 ml.) under reduced pressure. Cool the residual oil, add petroleum ether (b.p. 40–60°; 50 ml.) and stir until the mixture solidifies. Filter off the solid product and wash with petroleum ether (b.p. 40–60°; 3 × 30 ml.).

3-Phenylbenzo-2,1-isoxazole crystallizes from 50% aqueous ethanol (ca. 10 ml./g.) (decolourizing carbon) as colourless prisms, m.p. 53–54°.

Yield = 2·5–2·8 g. (81·5–91·5%).

## Reviews

(a) "Isoxazoles and Related Compounds", "The Chemistry of Heterocyclic Compounds", ed. A. Weissberger, Vol. 17, p. 166.
(b) "Isoxazoles", "Heterocyclic Compounds", ed. R. C. Elderfield, Vol. 5, p. 452.
(c) "Anthranils", "Chemistry of Carbon Compounds", ed. E. H. Rodd, Vol. IV A, p. 351.
(d) "Nitrenes", see review references on p. 58.

## References

1. P. A. S. Smith, B. B. Brown, R. K. Putney, and R. F. Reinisch, *J. Amer. Chem. Soc.*, 1953, **75**, 6335.
2. H. J. Scheifele jun., and D. F. DeTar, *Org. Synth.*, 1963, Coll. Vol. IV, 34.

## 2-METHYLBENZOXAZOLE

When *o*-aminophenols are heated with carboxylic acids, acid chlorides or anhydrides etc., 2-substituted benzoxazoles are produced.[1] *o*-Aminophenol and acetic anhydride yield 2-methylbenzoxazole (I).[2]

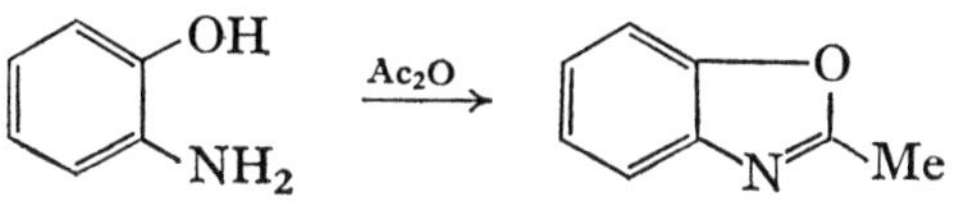


I

### 2-Methylbenzoxazole (I)

In a 100-ml. round-bottomed flask, cautiously mix *o*-aminophenol (10·9 g.) and acetic anhydride (31 g.; 28·5 ml.). Equip the flask with still-head and air condenser, etc., then heat the mixture gently until a uniform solution is formed and a colourless liquid begins to distil. Continue heating, collecting the distillate between 120 and 245° (there is virtually no residue). Using a short column, fractionate the distillate neglecting a fore-run (ca. 25 ml.) and a small residue.

2-Methylbenzoxazole is obtained as a colourless liquid, b.p. 200–208°.

Yield = 10–12 g. (75–90%).

## Spectra

(a) Ultraviolet.[3]
(b) Infrared.[4,5]

### Reviews

(a) "Benzoxazole and Related Systems", "Heterocyclic Compounds", ed. R. C. Elderfield, Vol. 5, p. 418.
(b) "Benzoxazoles","Chemistry of Carbon Compounds", ed. E. H. Rodd, Vol. IV A, p. 375.

### References

1. A. Ladenburg, *Ber.*, 1876, **9**, 1524.
2. W. Theilacker, *J. prakt. Chem.*, 1939, **153**, 54.
3. R. Passerini, *J. Chem. Soc.*, 1954, 2256.
4. D. G. O'Sullivan, *J. Chem. Soc.*, 1960, 3278.
5. P. Bassignana, C. Cogrossi, and M. Gandino, *Spectrochim. Acta*, 1963, **19**, 1885.

## 2-METHYL-6-NITROBENZOXAZOLE

Certain aromatic azides decompose thermally in a mixture of polyphosphoric acid (PPA) and a carboxylic acid to give benzoxazoles. The reaction proceeds via a nitrene intermediate and involves the diacyl derivative of the *o*-aminophenol.

Interaction of *p*-nitrophenylazide (I) (p. 49) with acetic and polyphosphoric acids leads to 2-methyl-6-nitrobenzoxazole (II).[1]

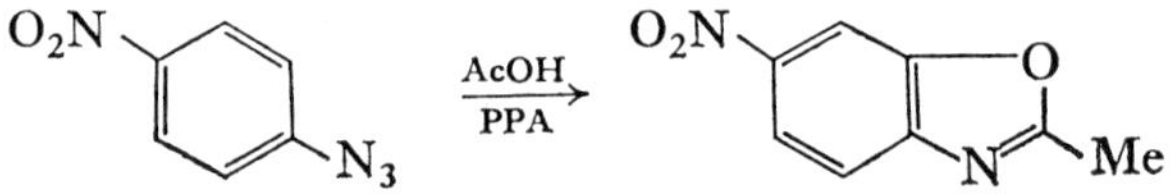


I II

### 2-Methyl-6-nitrobenzoxazole (II)

To mechanically stirred polyphosphoric acid (20 g.) at 80°, add dropwise over 20 min. a solution of *p*-nitrophenylazide (2 g.) in glacial acetic acid (20 ml.) (see apparatus on p. 38 and use carbon tetrachloride as heating liquid). When the addition is complete, replace the carbon tetrachloride by acetic anhydride (b.p. 138°), and stir the mixture at 138° for a further 2 hr. Pour the resulting solution into ice–water (100 ml.), and extract the mixture with chloroform (3 × 25 ml.). Combine the extracts, dry ($MgSO_4$), and evaporate off the chloroform to obtain the crude product as a buff solid.

2-Methyl-6-nitrobenzoxazole crystallizes from ethanol (ca. 10 ml./g.) as fine colourless needles, m.p. 157–158°.

Yield = 1·74–1·84 g. (80–85%).

### Spectra

(a) Ultraviolet.[2]
(b) Infrared.[3]

## Reviews

"Benzoxazoles", see review references on p. 56.

## References

1. R. Garner, E. B. Mullock, and H. Suschitzky, *J. Chem. Soc.*, 1966, 1980.
2. R. Passerini, *J. Chem. Soc.*, 1954, 2256.
3. D. G. O'Sullivan, *J. Chem. Soc.*, 1960, 3278.

# BENZO-2,1,3-OXADIAZOLE 1-OXIDE (BENZOFUROXAN)

Benzo-2,1,3-oxadiazole 1-oxide (II)[1,2] can be obtained in excellent yield from the decomposition of *o*-nitrophenylazide (I). Kinetic measurements[3] indicate that the reaction proceeds by a neighbouring-group displacement of nitrogen from the azide group by the nitro group, rather than by a nitrene intermediate.

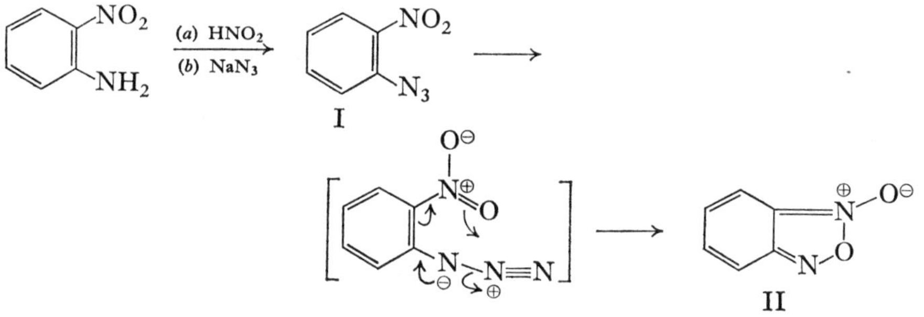


### *o*-Nitrophenylazide (I)

Add finely ground *o*-nitroaniline (6·9 g.) to a mixture of concentrated hydrochloric acid (25 ml.) and water (75 ml.), and stir the suspension for 5 min. Cool in an ice-bath, add a solution of sodium nitrite (3·8 g.) in water (20 ml.), and stir the mixture for 15 min. at 0°, after which time most of the nitroamine will have dissolved. Filter off any undissolved amine, and run the clear diazonium chloride solution into a stirred solution of sodium azide (5 g.) and sodium acetate (60 g.) in water (150 ml.).

#### Note

All operations involving sodium azide must be carried out in an efficient fume cupboard, since aqueous solutions of this compound evolve hydrazoic acid ($HN_3$) which is highly toxic.

Filter off the solid product, wash with cold water, and dry in a vacuum desiccator over concentrated sulphuric acid.

*o*-Nitrophenylazide is obtained as a pale yellow solid, m.p. 53–54°, sufficiently pure for direct use in the next stage.

Yield=7·5–8 g. (91·5–97·5%).

**Note**

Nearly all azides decompose violently when heated alone either in the solid or the liquid state. The prepared azide must *not*, therefore, be dried over a flame or in an oven.

### Benzo-2,1,3-oxadiazole 1-oxide (II)

To xylene (70 ml.) maintained at 100° in a 250-ml. round-bottomed flask, equipped with reflux condenser and dropping funnel, add dropwise over 15 min. a solution of *o*-nitrophenylazide (3·3 g.) in xylene (30 ml.). Heat the reaction mixture at 100° for 2 hr., then distil off the xylene under reduced pressure to a volume of ca. 10 ml. Cool, and stir the oily residue with petroleum ether (b.p. 40–60°; 60 ml.) until the product solidifies. Filter, and wash the residue with petroleum ether (b.p. 40–60°; 4 × 50 ml.).

Benzo-2,1,3-oxadiazole 1-oxide crystallizes from 50% aqueous ethanol (ca. 6 ml./g.) as pale-yellow prisms, m.p. 70–71°.

Yield=2·2–2·5 g. (81–92%).

**Spectra**

(a) Ultraviolet.[3]
(b) Infrared.[4]
(c) Proton magnetic resonance.[5]

**Reviews**

(a) "Nitrenes", L. Horner and A. Christmann, *Angew. Chem.* (*Internat. edn.*), 1963, **2**, 599.
(b) "Preparation and Properties of Imido Intermediates (Imidogens)", R. A. Abramovitch and B. A. Davis, *Chem. Rev.*, 1964, **64**, 149.
(c) "Oxadiazoles and Related Compounds", "The Chemistry of Heterocyclic Compounds", ed. A. Weissberger, Vol. 17, 235.
(d) "Oxadiazoles", "Heterocyclic Compounds", ed. R. C. Elderfield, Vol. 7, p. 479.

**References**


1. E. Noelting and K. Kohn, *Chem. Zentr.*, 1894, **18**, 1095.
2. P. A. S. Smith and J. H. Boyer, *Org. Synth.*, 1963, Coll. Vol. IV, 75.
3. E. A. Birkhimer, E. A. Norup, and T. A. Bak, *Acta. Chem. Scand.*, 1960, **14**, 1894.
4. R. J. Gaughran, J. P. Picard, and J. V. R. Kaufman, *J. Amer. Chem. Soc.*, 1954, **76**, 2233.
5. R. K. Harris, A. R. Katritzky, S. Øksne, A. S. Bailey, and W. G. Paterson, *J. Chem. Soc.*, 1963, 197.

## 5(6)-CHLOROBENZO-2,1,3-OXADIAZOLE 1-OXIDE (5(6)-CHLOROBENZOFUROXAN)

Benzo-2,1,3-oxadiazole 1-oxides are readily prepared from substituted *o*-nitrophenylazides,[1] and have recently been shown[2] to be tautomeric structures, the "5" and "6" positions being identical. 5(6)-Chlorobenzo-2,1,3-oxadiazole 1-oxide (IVa–IVb) may, therefore, be prepared either by the decomposition of 4-chloro-2-nitrophenylazide (III) as given below, or by the decomposition of 5-chloro-2-nitrophenylazide (V).

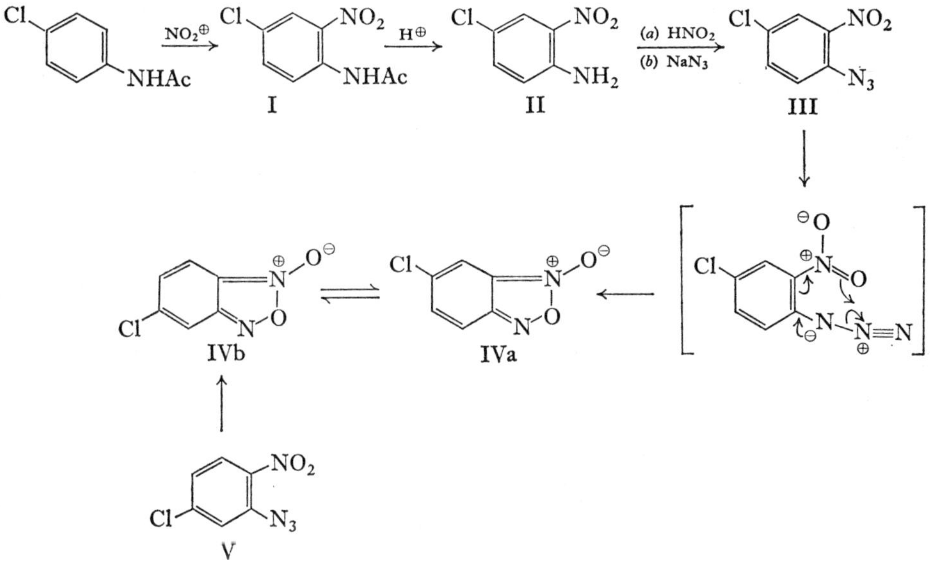


### 4-Chloro-2-nitroacetanilide (I)

Add *p*-chloroacetanilide (7·5 g.) in small portions to an ice-cold, stirred mixture of concentrated nitric acid (sp. gr. 1·42; 20 ml.) and concentrated sulphuric acid (20 ml.) contained in a 100 ml. conical flask. Stir the mixture at 0° for 10 min., then at room-temperature for a further 0·5 hr. Pour the solution onto crushed ice (200 g.), filter off the precipitated yellow solid and wash with cold water (3 × 50 ml.).

4-Chloro-2-nitroacetanilide crystallizes from ethanol (ca. 10 ml./g.) as pale yellow needles, m.p. 100–101°.

Yield = 7–7·5 g. (74–79%).

### 4-Chloro-2-nitroaniline (II)

Dissolve 4-chloro-2-nitroacetanilide (6 g.) in cold concentrated sulphuric acid (30 ml.) and heat the solution on a water-bath at 100° for 15 min. Cool the reaction mixture and cautiously stir into crushed ice (200 g.). Filter off the precipitated product and wash with cold water.

4-Chloro-2-nitroaniline crystallizes from water (ca. 10 ml./g.) as orange needles, m.p. 116°.

Yield = 4·2–4·5 g. (87–93%).

### 4-Chloro-2-nitrophenylazide (III)

Add finely ground 4-chloro-2-nitroaniline (3 g.) to a mixture of concentrated hydrochloric acid (20 ml.) and water (60 ml.). Stir the suspension for 5 min., then cool in an ice-bath and add a solution of sodium nitrite (2 g.) in water (10 ml.). Stir for 15 min. at 0°, then filter off any undissolved amine and run the clear diazonium chloride solution into a stirred solution of sodium azide (2·5 g.) and sodium acetate (50 g.) in water (150 ml.).

**Note**

All operations involving sodium azide must be carried out in an efficient fume cupboard, since aqueous solutions of this compound evolve hydrazoic acid ($HN_3$) which is highly toxic.

Filter off the solid product, wash with cold water and dry in a vacuum desiccator over concentrated sulphuric acid.

4-Chloro-2-nitrophenylazide is obtained as a pale yellow solid, m.p. 70–72°, sufficiently pure for direct use in the next stage.

Yield = 3·2–3·4 g. (92·5–98·5%).

**Note**

Nearly all azides decompose violently when heated alone either in the solid or the liquid state. The prepared azide must *not*, therefore, be dried over a flame or in an oven.

### 5(6)-Chlorobenzo-2,1,3-oxadiazole-1-oxide (IVa–IVb)

In a 250-ml. 2-necked round-bottomed flask, fitted with reflux condenser and dropping-funnel, heat toluene (75 ml.) under reflux. To the boiling solvent, add dropwise over 15 min., a solution of 4-chloro-2-nitrophenylazide (2 g.) in toluene (25 ml.). Heat the solution under reflux for 2 hr., then distil off the solvent to a volume of ca. 10 ml., and cool the residual oil. Add petroleum ether (b.p. 40–60°; 60 ml.), stir until solidification occurs and filter off the product.

5(6)-Chlorobenzo-2,1,3-oxadiazole 1-oxide crystallizes from 50% aqueous ethanol (ca. 20 ml./g.) as pale-yellow needles, m.p. 48°.

Yield = 1·4–1·5 g. (81·5–87·5%).

### Reviews

"Benzo-2,1,3-oxadiazole 1-oxides", see review references on p. 58.

### References


1. See references given in the review on imidogens (nitrenes) by R. A. Abramovitch and B. A. Davis, *Chem. Rev.*, 1964, **64**, 149.
2. A. R. Katritzky, S. Øksne, and R. K. Harris, *Chem. and Ind.*, 1961, 990; R. K. Harris, A. R. Katritzky, S. Øksne, A. S. Bailey, and W. G. Paterson, *J. Chem. Soc.*, 1963, 197.


## 2-PHENYLBENZO-1,3-DIAZA-2-BOROLE

Diazaboroles result from condensation of *o*-diamines with phenylboron dichloride. Interaction of *o*-phenylenediamine with phenylboron dichloride yields 2-phenylbenzo-1,3-diaza-2-borole (I).

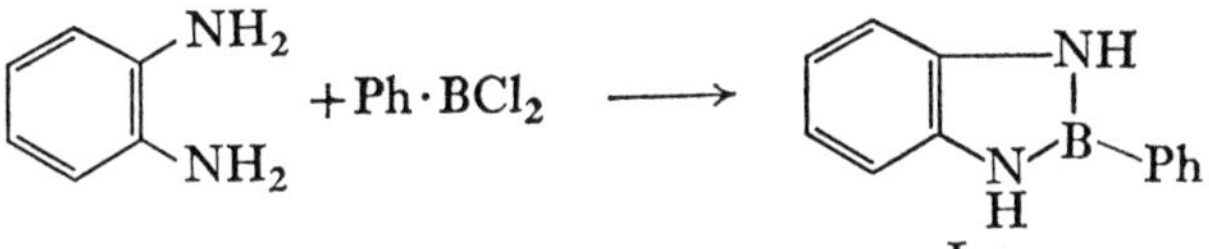


I

### 2-Phenylbenzo-1,3-diaza-2-borole (I)

Using a dropping pipette, add, as rapidly as possible, phenylboron dichloride (3·2 g.) to sodium-dried benzene (50 ml.). Add this solution dropwise to a solution of *o*-phenylenediamine (2·2 g.) in sodium-dried benzene (100 ml.) contained in a 250-ml. 2-necked, round-bottomed flask. Heat the mixture under reflux on a water-bath at 100°, until hydrogen chloride evolution ceases (ca. 6 hr.). Filter the hot mixture, and evaporate the filtrate to obtain the crude product as a brown solid.

2-Phenylbenzo-1,3-diaza-2-borole crystallizes from petroleum ether (b.p. 100–120°; ca. 30 ml./g.) as white plates, m.p. 204°.

Yield = 3–3·6 g. (77·5–93%).

### Spectra

(a) Ultraviolet.[1]

(b) Infrared.[2]

## Reviews

(a) "Heterocyclic Organic Boron Compounds", P. J. Maitlis, *Chem. Rev.*, 1962, **62**, 223.
(b) "Studies in Boron Chemistry; Organic Analogues of Heterocycles", W. Gerrard, *Chem. and Ind.*, 1966, 832.
(c) "Boron-Nitrogen Heterocycles", D. A. Payne and E. A. Eads, *J. Chem. Educ.*, 1964, **41**, 334.
(d) "Heteroaromatic Boron Compounds", M. J. S. Dewar, *Progr. Boron Chem.*, 1964, **1**, 235.

## References

1. M. J. S. Dewar, V. P. Kubba, and R. Pettit, *J. Chem. Soc.*, 1958, 3076.
2. F. K. Butcher, W. Gerrard, M. Howarth, E. F. Mooney, and H. A. Willis, *Spectrochim. Acta*, 1964, **20**, 79.

## 2-PHENYLBENZO-1,3-DIAZA-2-PHOSPHOLE 2-OXIDE

Benzo-1,3-diaza-2-phospholes are prepared by reacting *o*-diamines with phosphorus dihalides. Interaction of *o*-phenylenediamine (I) with phenyl phosphonic dichloride (II) in an inert solvent yields 2-phenylbenzo-1,3-diaza-2-phosphole 2-oxide (III).[1]

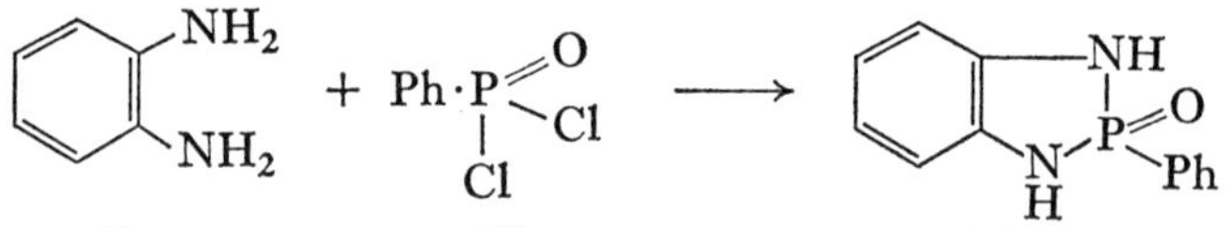


I II III

### 2-Phenylbenzo-1,3-diaza-2-phosphole 2-oxide (III)

In a 500-ml. 2-necked, round-bottomed flask fitted with dropping-funnel and reflux condenser, magnetically stir a solution of *o*-phenylenediamine (2·7 g.) in bromobenzene (100 ml.). To the solution add dropwise over 15 min. a solution of phenyl phosphonic dichloride (4·8 g.) in bromobenzene (10 ml.). Heat the mixture in an oil-bath at 130° for 0·5 hr., then under reflux for a further 10 min. Cool the solution, filter off the precipitated product and wash the residue with petroleum ether (b.p. 40–60°; 3 × 50 ml.).

2-Phenylbenzo-1,3-diaza-2-phosphole 2-oxide crystallizes from bromobenzene (ca. 60 ml./g.) as off-white elongated prisms, m.p. 276–278°.

Yield = 4·5–5·2 g. (78–90%).

**Spectra**

Infrared.[2]

**Reviews**

(a) "Heterocyclic Derivatives of Phosphorus, Arsenic and Antimony", F. G. Mann, *Progr. Org. Chem.*, 1958, **4**, 217.

(b) " Heterocyclic Derivatives of Phosphorus", "The Chemistry of Heterocyclic Compounds", ed. A. Weissberger, Vol. 1, p. 9.

(c) "Five-membered Rings Containing Phosphorus and Oxygen and Nitrogen", "Chemistry of Carbon Compounds", ed. E. H. Rodd, Vol. IV B, p. 1170.

**References**

1. R. A. Dannley and P. L. Wagner, *J. Org. Chem.*, 1961, **26**, 3995.
2. V. Gutmann, D. E. Hagen, and K. Utvary, *Monatsh.*, 1962, **93**, 627.

## 2-PHENYLBENZO-1,3-OXAZA-2-BOROLE

Oxazaboroles result from the condensation of *o*-aminophenols with phenylboronic acid. Interaction of *o*-aminophenol with phenylboronic acid (I) yields 2-phenylbenzo-1,3-oxaza-2-borole (II).[1]

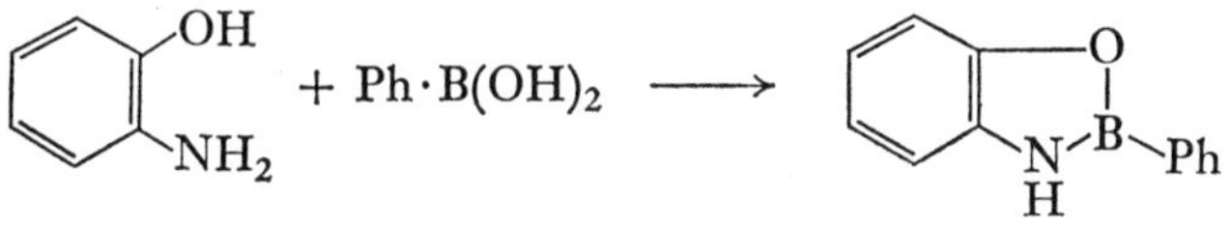


I II

### 2-Phenylbenzo-1,3-oxaza-2-borole (II)

In a 500-ml. round-bottomed flask, fitted with a Dean and Stark apparatus,[2] heat under reflux a mixture of *o*-aminophenol (2·18 g.) and phenylboronic acid (2·4 g.) in dry toluene (250 ml.). When the theoretical quantity of water (0·75 ml.) is obtained (after ca. 6 hr.), filter the hot solution and evaporate off the toluene under reduced pressure to obtain the crude product as a brown solid.

2-Phenylbenzo-1,3-oxaza-2-borole crystallizes from benzene (ca. 10 ml./g.), (decolourizing carbon), as white plates, m.p. 105°.

Yield = 2·7–3 g. (69·5–77%).

**Spectra**

Ultraviolet.[3]

**Reviews**

"Boron Heterocycles", see review references on p. 62.

### References

1. M. Pailer and W. Fenzl., *Monatsh.*, 1961, **92**, 1294.
2. A. I. Vogel, "A Textbook of Practical Organic Chemistry", Longmans, Green & Co., London, 3rd Edn., 1956, p. 429.
3. M. J. S. Dewar, V. P. Kubba, and R. Pettit, *J. Chem. Soc.*, 1958, 3076.

## 2-PHENYLBENZOTHIAZOLE

On treatment with sodium disulphide, *o*-nitrochlorobenzene (I) yields 2,2′-dinitrodiphenyldisulphide (II),[1] which undergoes reductive cleavage when heated with zinc and acetic acid, the product being zinc *o*-aminobenzenethiol (III).[2] 2-Phenylbenzothiazole (IV) results from the action of benzoyl chloride on the latter.[2]

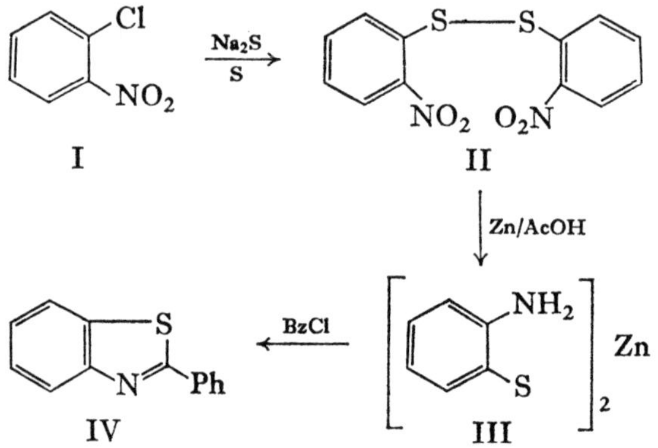


### 2,2′-Dinitrodiphenyldisulphide (II)

Heat a mixture of sodium sulphide (7·6 g.) and sulphur (1 g.) in water (10 ml.) until solution is complete (ca. 5 min.). Add this solution to a solution of *o*-nitrochlorobenzene (9·45 g.) in ethanol (12 ml.) in a 100-ml. 2-necked flask at such a rate that the flask contents reflux gently (addition takes ca. 0·5 hr.). When addition is complete, continue to heat under reflux for 2·5 hr., then filter the hot mixture and wash the residue with hot water. Dry at 60°.

2,2′-Dinitrodiphenyldisulphide is obtained as a waxy yellow solid, m.p. 173–175°, sufficiently pure for direct use in the next stage.

Yield = 5·5–6·5 g. (60–70%).

### Zinc *o*-aminobenzenethiol (zinc salt of *o*-aminothiophenol) (III)

To a solution of 2,2′-dinitrodiphenyldisulphide (2·6 g.) in warm glacial acetic acid (250 ml.) in a 500-ml. 2-necked, round-bottomed flask, add zinc dust (17 g.)

portionwise over 0·5 hr. (Exothermic reaction!) When the addition is complete, heat the mixture under reflux for 0·5 hr., then filter to remove any excess of zinc. Heat the filtrate to 100°, add boiling water (500 ml.) and allow to cool.

Zinc *o*-aminobenzenethiol is obtained as an off-white microcrystalline solid.

Yield = 1·6–1·7 g. (60–64%).

## 2-Phenylbenzothiazole (IV)

In a 10-ml. pear-shaped flask, heat a mixture of zinc *o*-aminobenzenethiol (1·6 g.) and benzoyl chloride (1·5 g.; 1·25 ml.) under reflux for 0·5 hr. (Heating should be controlled in order to avoid excessive charring). Allow to cool and grind up the crude product with warm 2N sodium hydroxide (15 ml.), then filter and wash the residue with water (15 ml.).

2-Phenylbenzothiazole crystallizes from 70% aqueous ethanol (ca. 10 ml./g.) as off-white needles, m.p. 113–114°.

Yield = 1·5–1·8 g. (70–84%).

### Spectra

Ultraviolet.[3]

### Reviews

(a) "Benzothiazoles", "Heterocyclic Compounds", ed. R. C. Elderfield, Vol. 5, p. 506.
(b) "Benzothiazoles", "Chemistry of Carbon Compounds", ed. E. H. Rodd, Vol. IV A, p. 425.

### References

1. T. Wohlfahrt, *J. prakt. Chem.*, 1902, **66**, 553.
2. M. T. Bogert and F. D. Snell, *J. Amer. Chem. Soc.*, 1924, **46**, 1308.
3. A. Cerniani and R. Passerini, *J. Chem. Soc.*, 1954, 2261.

## BENZO-2,1,3-THIADIAZOLE

Benzo-2,1,3-thiadiazoles result from condensation of *o*-diamines with thionyl chloride. Interaction of *o*-phenylenediamine with thionyl chloride yields benzo-2,1,3-thiadiazole (I).[1,2]

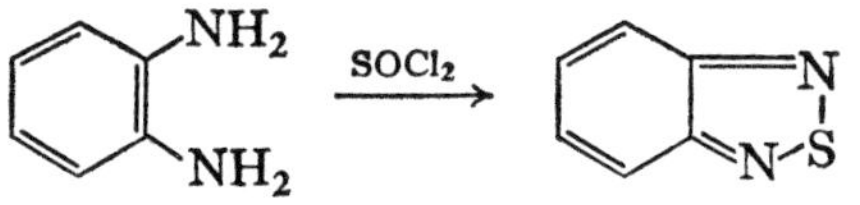


I

### Benzo-2,1,3-thiadiazole (I)

To a solution of *o*-phenylenediamine (10 g.) in dry toluene (200 ml.), contained in a 500-ml. round-bottomed flask, add cautiously freshly distilled thionyl chloride (30 ml.). Fit the flask with an efficient condenser and heat the mixture under reflux overnight. (The initially formed white precipitate should disappear after ca. 2 hr., leaving a deep-red solution.) Distil off the solvent together with the excess of thionyl chloride (fraction b.p. 90–120°; 220–225 ml.), and then distil the residual oil at atmospheric pressure.

Benzo-2,1,3-thiadiazole is obtained as a yellow oil, b.p. 210–215°, which readily solidifies to a crystalline solid, m.p. 42°.

Yield = 6·5–7·5 g. (51·5–59·5%).

## Spectra

Ultraviolet.[3]

## Reviews

(a) "The Thiadiazoles", "Heterocyclic Compounds", ed. R. C. Elderfield, Vol. 7, p. 581.

(b) "1,2,5-Thiadiazoles", "Chemistry of Carbon Compounds", ed. E. H. Rodd, Vol. IV A, p. 474.

## References

1. O. Hinsberg, *Ber.*, 1889, **22**, 2895.
2. A. M. Khaletskii and V. G. Pesin, *J. Gen. Chem.* (*U.S.S.R.*), 1950, **20**, 1914.
3. E. Sawicki and A. Carr, *J. Org. Chem.*, 1957, **22**, 503.

# 5 | SIX-MEMBERED SYSTEMS CONTAINING ONE HETEROATOM

## 3-CHLORO-2,6-DIMETHYLPYRIDINE

Thermal decomposition of sodium trichloroacetate yields the reactive intermediate dichlorocarbene (I). In the presence of dichlorocarbene, 2,5-dimethylpyrrole (II) (p. 2) undergoes ring expansion, yielding 3-chloro-2,6-dimethylpyridine (III).[1]

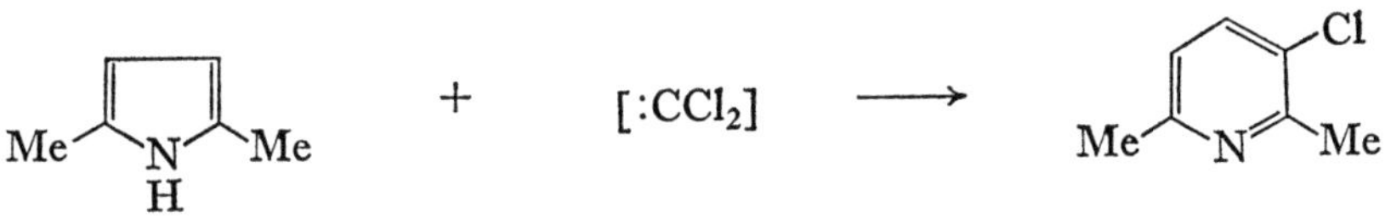


II I III

(from decomposition of sodium trichloroacetate)

### Sodium trichloroacetate

In a 250-ml. 2-necked, round-bottomed flask, prepare a solution of trichloroacetic acid (32·8 g.) in dry methanol (30 ml.). Cool the flask and contents in an ice–water bath, and rapidly add a cooled solution of sodium (4·6 g.) in dry methanol (80 ml.). Evaporate the resulting solution to dryness under reduced pressure at a temperature not exceeding 40°.

Sodium trichloroacetate is obtained as a white flaky crystalline solid and when completely dry, can be used directly in the next stage.

### 3-Chloro-2,6-dimethylpyridine (III)

To the sodium trichloroacetate (ca. 58 g.) contained in the 250-ml. flask from the previous experiment, add 2,5-dimethylpyrrole (9·5 g.; 10·2 ml.) and dry

1,2-dimethoxyethane (100 ml.). Heat the mixture under reflux in a dry nitrogen atmosphere for 24 hr., then cool and filter off the inorganic material. Distil off the solvent from the filtrate, and after cooling, dissolve the residue in ice-cold 2N hydrochloric acid (120 ml.). Thoroughly wash the resulting solution with ether (6 × 25 ml.) to remove unchanged pyrrole, and discard the ethereal extracts. Basify the aqueous solution with 4N sodium hydroxide (ca. 15 ml.), filter off any precipitate that may have formed, and extract the filtrate with ether (3 × 40 ml.). Combine the extracts, dry ($MgSO_4$), and after evaporating off the ether, distil the residue under water-pump vacuum.

3-Chloro-2,6-dimethylpyridine is obtained as a colourless liquid, b.p. 65–69°/11 mm.

Yield = 6–9 g. (42–64%).

## Spectra

(a) Ultraviolet.[1]
(b) Infrared.[1]
(c) Proton magnetic resonance.[1]

## Reviews

(a) "Pyridines", see review references on p. 70.
(b) "Carbenes, Reactive Intermediates Containing Divalent Carbon", E. Chinoporos, *Chem. Rev.*, 1963, **63**, 235.
(c) "Carbene Chemistry", W. Kirmse, Academic Press Inc., New York, London, 1964.
(d) "The Reactions of Heterocyclic Compounds with Carbenes", *Adv. Heterocyclic Chem.*, 1964, **3**, 57.

## Reference

1. R. L. Jones, Ph.D. thesis, University of Leicester.

# 3,5-DIETHOXYCARBONYL-2,6-DIMETHYL-4-(*m*-NITROPHENYL)PYRIDINE

The Hantzsch dihydropyridine synthesis involves interaction of a $\beta$-keto-ester with an aldehyde and ammonia.[1] The resulting dihydropyridine is readily oxidized to the pyridine. In this example, ethyl acetoacetate, *m*-nitrobenzaldehyde and ammonia condense together to form 3,5-diethoxycarbonyl-1,4-dihydro-2,6-dimethyl-4-(*m*-nitrophenyl)pyridine (I)[2] which is oxidized by nitric acid to 3,5-diethoxycarbonyl-2,6-dimethyl-4-(*m*-nitrophenyl)pyridine (II).

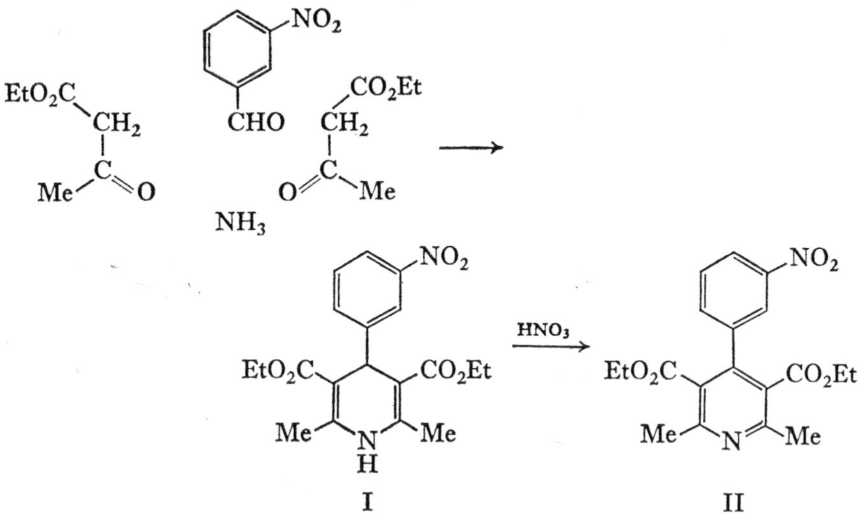

### 3,5-Diethoxycarbonyl-1,4-dihydro-2,6-dimethyl-4-(*m*-nitrophenyl)-pyridine (I)

In a 250-ml. round-bottomed flask, heat a solution of *m*-nitrobenzaldehyde (15·1 g.), ethyl acetoacetate (28·6 g.; 28 ml.), and concentrated ammonium hydroxide (8 ml.) in ethanol (60 ml.) under reflux for 3 hr. To the resulting mixture, add warm water (40 ml.) then allow to cool. Filter off the separated product and wash with 60% aqueous ethanol (10 ml.).

3,5-Diethoxycarbonyl-1,4-dihydro-2,6-dimethyl-4-(*m*-nitrophenyl)pyridine is obtained as yellow prisms, m.p. 165–167°, sufficiently pure for direct use in the next stage.

Yield = 16–18 g. (43–48%).

### 3,5-Diethoxycarbonyl-2,6-dimethyl-4-(*m*-nitrophenyl)pyridine (II)

In a 250-ml. conical flask, heat a solution of 3,5-diethoxycarbonyl-1,4-dihydro-2,6-dimethyl-4-(*m*-nitrophenyl)pyridine (10 g.) in 4N nitric acid (150 ml.) on the water-bath at 100° for 1 hr. Cool, and basify the mixture with 30% aqueous sodium hydroxide, then extract the liberated oil with ether (3 × 50 ml.). Combine the extracts, dry ($MgSO_4$), and after evaporating off the ether, distil the residue under oil-pump vacuum.

3,5-Diethoxycarbonyl-2,6-dimethyl-4-(*m*-nitrophenyl)pyridine is obtained as an almost colourless oil, b.p. 187–188°/1·5 mm., which readily solidifies, m.p. 60–62°.

Yield = 7·5–8·5 g. (75–85%).

**Spectra**

(a) Ultraviolet.[3]

## Reviews

(a) "Pyridine and its Derivatives", "The Chemistry of Heterocyclic Compounds", ed. A. Weissberger, Vol. 14 (parts 1–4).
(b) "The Chemistry of the Pyridines", "Heterocyclic Compounds", ed. R. C. Elderfield, Vol. I, p. 397.
(c) "Pyridine and its Derivatives", "Chemistry of Carbon Compounds", ed. E. H. Rodd, Vol. IV A, p. 488.

## References

1. A. Hantzsch, *Annalen*, 1882, **215**, 1.
2. A. P. Phillips, *J. Amer. Chem. Soc.*, 1949, **71**, 4003.
3. A. Kamal, M. Ahmad, N. Mond, and A. M. Hamid, *Bull. Chem. Soc. Japan*, 1964, **37**, 610.

# 4-NITROPYRIDINE *N*-OXIDE

Pyridine is extremely resistant towards electrophilic attack. Formation of the *N*-oxide, however, facilitates nitration of the pyridine nucleus mainly at the 4-position.[1] Such a reaction can be understood by referring to the major canonical forms of pyridine *N*-oxide (I–III).[2]

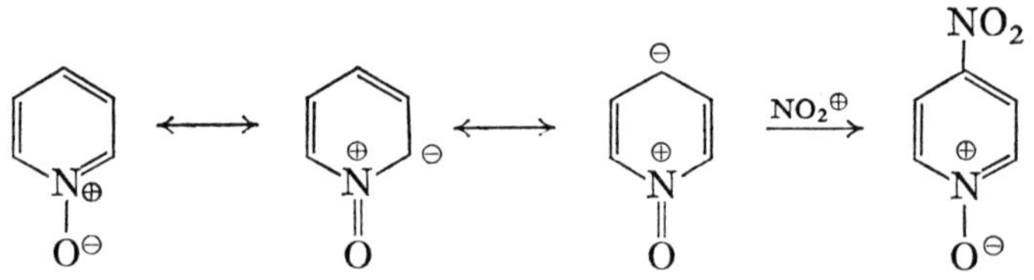

I II III IV

### 4-Nitropyridine *N*-oxide (IV)

Dissolve dry pyridine *N*-oxide[3] (10 g.) in a mixture of fuming nitric acid (sp. gr. 1·52; 10 ml.) and concentrated sulphuric acid (30 ml.), contained in a 250-ml. 2-necked, round-bottomed flask, fitted with an internal thermometer. Heat the solution for 3·5 hr. at 128–130° in an oil-bath. Cool the reaction mixture, pour onto crushed ice (250 g.), and neutralize the resulting solution with concentrated ammonium hydroxide. Filter off the pale-yellow product, and dry at 100°. Extract the filtrate with chloroform (2 × 30 ml.), combine the extracts, dry ($MgSO_4$), and evaporate off the chloroform to obtain further material as a pale-yellow residue. Combine this residue with that obtained by filtration, triturate with ether (100 ml.), and filter off the product.

4-Nitropyridine *N*-oxide crystallizes from acetone (ca. 5 ml./g.) as pale-yellow prisms, m.p. 158–159°.

Yield = 9·8–10·2 g. (66·5–69%).

### Note

If during nitration, the temperature is allowed to rise above 130°, the product will be contaminated with significant amounts of 2-nitro- and 4-nitropyridine.

### Spectra

(a) Ultraviolet.[4]
(b) Infrared.[5]
(c) Proton magnetic resonance.[6]

### Reviews

(a) "Aromatic Heterocyclic *N*-oxides", D. V. Ioffe and L. S. Efros, *Russ. Chem. Rev.*, 1961, **30**, 569.
(b) "The Chemistry of the Aromatic Heterocyclic *N*-oxides", A. R. Katritzky, *Quart. Rev.*, 1956, **10**, 395.
(c) "Pyridine *N*-oxide", "The Chemistry of Heterocyclic Compounds", ed. A. Weissberger, Vol. 14, (part 2), p. 97.

### References

1. E. Ochiai, *J. Org. Chem.*, 1953, **18**, 534.
2. E. P. Linton, *J. Amer. Chem. Soc.*, 1940, **62**, 1945.
3. H. S. Mosher, L. Turner, and A. Carlsmith, *Org. Synth.*, 1963, Coll. Vol. IV, 828.
4. H. H. Jaffé, *J. Amer. Chem. Soc.*, 1955, **77**, 4451.
5. A. R. Katritzky and J. N. Gardner, *J. Chem. Soc.*, 1958, 2192.
6. A. R. Katritzky and J. M. Lagowski, *J. Chem. Soc.*, 1961, 43.

## 3-CYANO-4,6-DIMETHYL-2-PYRIDONE (METHOD 1)

In this Guareschi-Thorpe[1] reaction, acetylacetone (I) and cyanoacetamide (II) condense under basic conditions to yield 3-cyano-4,6-dimethyl-2-pyridone (III).[2] (See also p. 72).

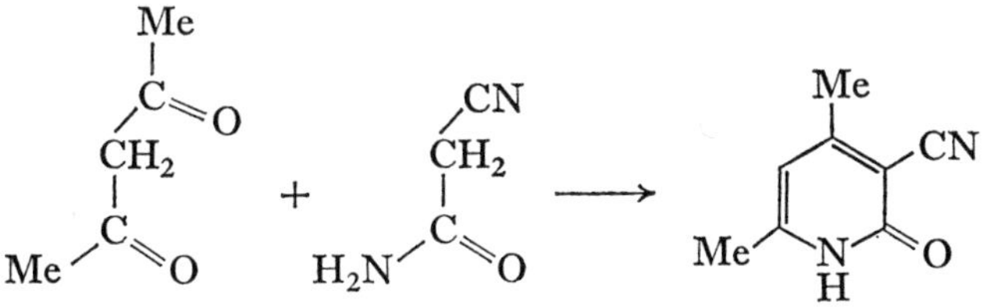


I II III

### 3-Cyano-4,6-dimethyl-2-pyridone (III)

To a solution of acetylacetone (4·85 g.; 5 ml.) and cyanoacetamide (4·5 g.) in 90% aqueous ethanol (35 ml.) in a 100-ml. round-bottomed flask, add piperidine (2 g.; 2·3 ml.). Heat the mixture under reflux for 0·5 hr., then cool and filter off the crystalline product. Wash the residue with ethanol.

3-Cyano-4,6-dimethyl-2-pyridone crystallizes from ethanol (ca. 100 ml./g.) as lustrous needles, m.p. 298–300°.

Yield = 5–5·5 g. (63–69%).

For a discussion on the tautomerism of pyridones, see ref. 3.

### Reviews

"Pyridines", see review references on p. 70.

### References

1. I. Guareschi, *Chem. Zentr.*, 1899, **1**, 289.
2. H. M. Van Wagtendonk and J. W. Wibaut, *Rec. Trav. chim.*, 1942, **61**, 728.
3. A. R. Katritzky and J. M. Lagowski, *Adv. Heterocyclic Chem.*, 1963, **1**, 347.

## 3-CYANO-4,6-DIMETHYL-2-PYRIDONE (METHOD 2) AND 3,5-DIACETYL-1,4-DIHYDRO-2,4,6-TRIMETHYLPYRIDINE

### ORGANIC REACTIONS IN AQUEOUS SOLUTION AT ROOM TEMPERATURE[1]

Water at room temperature is an effective medium for simple condensations leading to pyrimidine, pyrrole, and pyridine derivatives etc., and such reactions are usually termed "syntheses under physiological (or 'cell-possible') conditions."[1]

*The first example* is a Guareschi–Thorpe pyridone synthesis.[2] Cyanoacetamide (I) and acetylacetone (II) condense together in the presence of aqueous potassium carbonate to yield 3-cyano-4,6-dimethyl-2-pyridone (III) (see also p. 71).

### Example 1

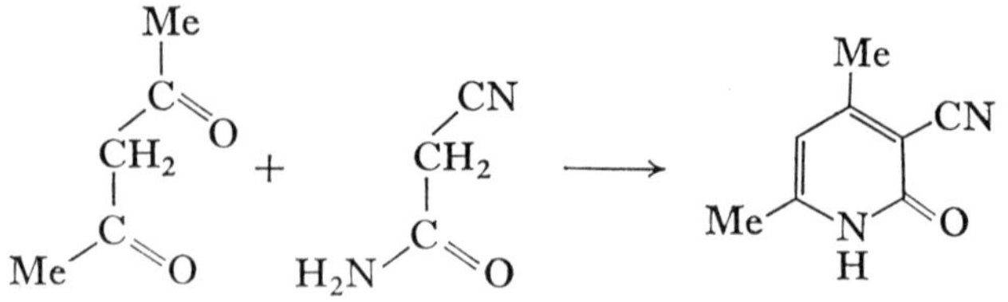


I II III

### 3-Cyano-4,6-dimethyl-2-pyridone (III)

In a 100-ml. conical flask, place cyanoacetamide (1·68 g.) and acetylacetone (2 g.; 2·05 ml.), and add ethanol (1 ml.) and a solution of potassium carbonate (0·8 g.) in water (40 ml.). Stopper the flask, and allow the mixture to stand at room temperature for 24 hr. Filter off the product, and wash with 90% aqueous ethanol (10 ml.).

3-Cyano-4,6-dimethyl-2-pyridone is obtained as white needles, m.p. 288–290°.

Yield = 2·5–2·8 g. (85–95%).

For a discussion on the tautomerism of pyridones, see ref. 3 on p. 72.

*The second example* is a Hantzsch dihydropyridine synthesis.[3] Acetaldehyde and acetylacetone condense together in the presence of aqueous ammonium carbonate to yield 3,5-diacetyl-1,4-dihydro-2,4,6-trimethylpyridine (IV).

**Example 2**

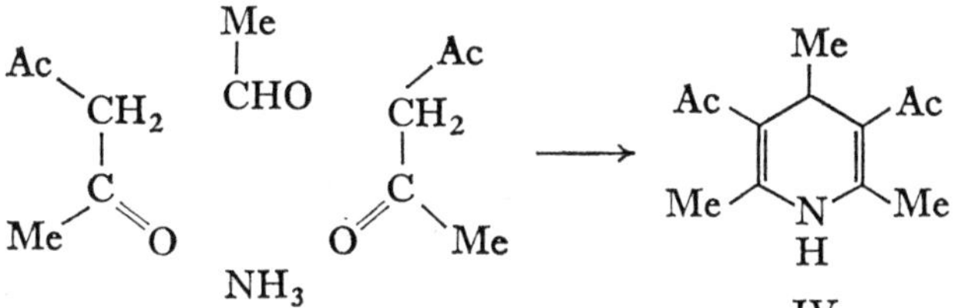


IV

### 3,5-Diacetyl-1,4-dihydro-2,4,6-trimethylpyridine (IV)

In a 50-ml. conical flask, dissolve ammonium carbonate (2 g.) in water (20 ml.), and to the solution add ethanol (1 ml.), acetylacetone (2·2 ml.) and acetaldehyde (0·6 ml.). Stopper the flask, and allow the mixture to stand at room temperature for 4 days. Filter off the product and wash with 10% aqueous ethanol (5 ml.).

3,5-Diacetyl-1,4-dihydro-2,4,6-trimethylpyridine is obtained as long yellow needles, m.p. 152–153°.

Yield = 0·8–1 g. (36–45%).

**Reviews**

"Pyridines", see review references on p. 70.

**References**


1. C. A. C. Haley and P. Maitland, *J. Chem. Soc.*, 1951, 3155.
2. I. Guareschi, *Chem. Zentr.*, 1899, **1**, 289.
3. A. Hantzsch, *Annalen*, 1882, **215**, 1.

# 4-PYRIDONE-2,6-DICARBOXYLIC ACID (CHELIDAMIC ACID)

γ-Pyrones are susceptible towards nucleophilic attack at the 2-position. If ammonia is used, the attack is accompanied by ring opening, followed by loss of water and subsequent cyclization to yield a pyridine derivative. γ-Pyrone-2,6-dicarboxylic acid (chelidonic acid)[1] (I) reacts[2] with concentrated ammonium hydroxide to give 4-pyridone-2,6-dicarboxylic acid (II).

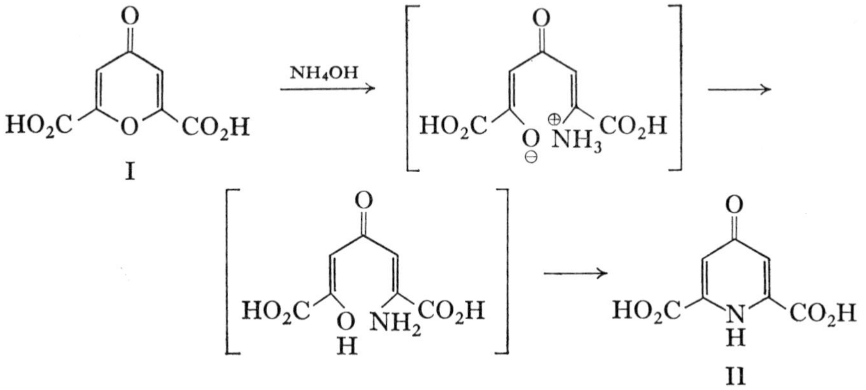

## 4-Pyridone-2,6-dicarboxylic acid (II)

In a 250-ml. round-bottomed flask, heat a solution of γ-pyrone-2,6-dicarboxylic acid (10 g.) in concentrated ammonium hydroxide (110 ml.) under reflux for 3 hr. Cool, and dilute the reaction mixture with cold water (200 ml.), then acidify with concentrated hydrochloric acid. Filter off the precipitate and wash with cold water.

4-Pyridone-2,6-dicarboxylic acid crystallizes from water (ca. 12 ml./g.) as white prisms, m.p. 250–255° (decomp.).

Yield = 8·1–8·6 g. (81·5–86·5%).

### Spectra

Ultraviolet.[3]

### Reviews

(a) "Pyridine Carboxylic Acids", "The Chemistry of Heterocyclic Compounds", ed. A. Weissberger, Vol. 14 (part 3), p. 179.
(b) "Pyridine Carboxylic Acids", "Heterocyclic Compounds", ed. R. C. Elderfield, Vol. I, p. 563.

(c) "Pyridine Carboxylic Acids", "Chemistry of Carbon Compounds", ed. E. H. Rodd, Vol. IV A, p. 562.

## References

1. E. R. Riegel and F. Zwilgmeyer, *Org. Synth.*, 1943, Coll. Vol. II, 126.
2. E. R. Riegel and M. C. Reinhard, *J. Amer. Chem. Soc.*, 1926, **48**, 1334.
3. L. R. Fibel and P. E. Spoerri, *J. Amer. Chem. Soc.*, 1948, **70**, 3908.

## 3-ACETYL-3,4-DIHYDRO-6-METHYL-4-OXO-α-PYRONE (DEHYDROACETIC ACID)

In the presence of base, diketen (I) yields mainly 3-acetyl-3,4-dihydro-6-methyl-4-oxo-α-pyrone (II).[1]

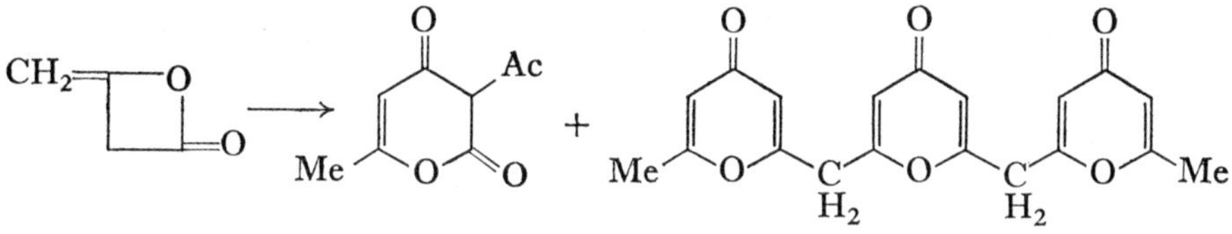

I II III

### 3-Acetyl-3,4-dihydro-6-methyl-4-oxo-α-pyrone (II)

In a 100-ml. 2-necked flask, fitted with magnetic stirrer, reflux condenser, and dropping-funnel, stir a suspension of sodium phenoxide (0·1 g.) in dry benzene (25 ml.). Heat the mixture to ca. 70°, and carefully add diketen ( 9 g.; 8·3 ml.) dropwise over 1 hr. When the addition is complete, heat the mixture under reflux for 1 hr., then remove benzene (18 ml.) by distillation, and allow the residual solution to stand overnight.

Filter off the crystalline product, and wash with a little benzene. (Obtain additional crude material by concentration of the mother liquor.)

Purify the crude product by dissolving it in hot methanol (ca. 3 ml./g.) and by hot filtration to remove the insoluble 2,6-bis-(6-methyl-4-oxo-2-pyranylmethyl)-pyrone (III), m.p. 235–236°.

On cooling, 3-acetyl-3,4-dihydro-6-methyl-4-oxo-α-pyrone is obtained as pale-buff needles, m.p. 109–110°.

Yield = 3·5–4·0 g. (39–45%).

## Spectra

(a) Ultraviolet.[2]
(b) Infrared.[3]
(c) Proton magnetic resonance.[4]

### Reviews

(a) "α- and γ-Pyrones", see review references on p. 77.
(b) "Diketen", see review reference on p. 97.

### References

1. A. B. Steel, A. B. Boese, and M. F. Dull, *J. Org. Chem.*, 1949, **14**, 460.
2. J. A. Berson, *J. Amer. Chem. Soc.*, 1952, **74**, 5172.
3. S. W. Nakhare and S. S. Deshapande, *Vikram, J. Vikram Univ.*, 1961, **5**, 24. (*Chem. Abs.*, 1963, **59**, 6348).
4. S. Forsen and M. Nilsson, *Arkiv Kemi*, 1961, **17**, 523.

## 2,6-DIMETHYL-γ-PYRONE

On heating in polyphosphoric acid (PPA), certain carboxylic acids or their anhydrides yield γ-pyrones. Under these conditions, acetic anhydride yields 2,6-dimethyl-γ-pyrone (I).[1]

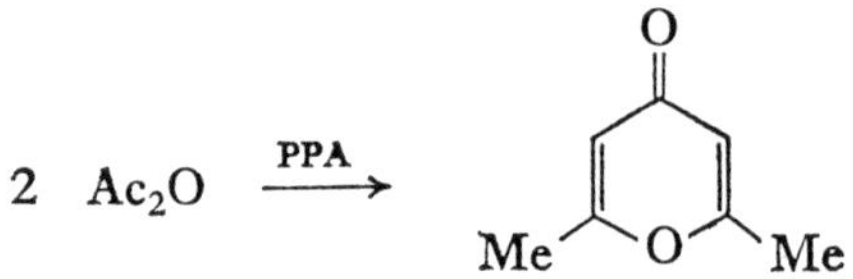


I

### 2,6-Dimethyl-γ-pyrone (I)

In a 100-ml. 3-necked, round-bottomed flask equipped with mechanical stirrer, internal thermometer and reflux condenser, stir a mixture of acetic anhydride (5 g.; 4·6 ml.) and polyphosphoric acid (50 g.) at 200–210° for 3 hr. Cool, and add the solution to ice–water (200 ml.). Repeatedly extract the resulting dark mass with chloroform and allow the combined extracts to stand over a mixture of dried magnesium sulphate and decolourizing carbon for 15 min. Filter, and evaporate off the chloroform to obtain the crude product as a buff-coloured solid.

2,6-Dimethyl-γ-pyrone sublimes under oil-pump vacuum (at ca. 100°/0·1 mm.) as white needles, m.p. 131–132°.

Yield = 1·67–1·97 g. (55–65%).

### Spectra

(a) Ultraviolet.[2]
(b) Infrared.[2]
(c) Proton magnetic resonance.[2]

## Reviews

(a) "γ-Pyrones", "Heterocyclic Compounds", ed. R. C. Elderfield, Vol. I, p. 370.
(b) "γ-Pyrones", "Chemistry of Carbon Compounds", ed. E. H. Rodd, Vol. IV B, p. 821.
(c) "The Chemistry of Monocyclic α- and γ-Pyrones", L. F. Cavalieri, *Chem. Rev.*, 1947, **41**, 525.
(d) "Polyphosphoric Acid", see review references on p. 78.

## References

1. E. B. Mullock and H. Suschitzky, *J. Chem. Soc.*, 1967, 828.
2. B. G. Engel, W. Brzeski, and P. A. Plattner, *Helv. Chim. Acta*, 1949, **32**, 1752.

## 6-ISOBUTENYL-4-METHYL-α-PYRONE

In the presence of polyphosphoric acid (PPA), 3,3-dimethylacrylic acid (I) undergoes self-condensation and cyclization to yield 6-isobutenyl-4-methyl-α-pyrone (II).[1]

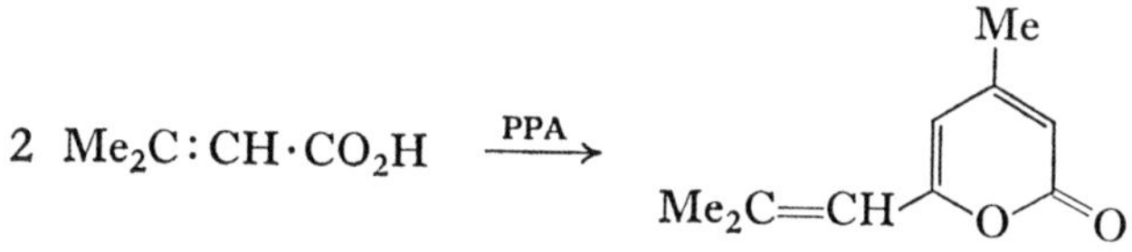


I II

### 6-Isobutenyl-4-methyl-α-pyrone (II)

Mechanically stir a mixture of 3,3-dimethylacrylic acid (10 g.) and polyphosphoric acid (120 g.) in a 250-ml. flask on a water-bath at 100° for 2 hr. Decompose the resulting orange complex by adding it to ice–water (750 g.) and extract the mixture with ether (3 × 50 ml.). Combine the extracts and wash the ethereal solution successively with water and aqueous sodium hydrogen carbonate to remove any unchanged acid. Dry the ethereal solution ($MgSO_4$), and after evaporating off the ether, distil the residual brown oil under water-pump vacuum.

6-Isobutenyl-4-methyl-α-pyrone is obtained as a pale-yellow oil, b.p. 160–164°/11 mm., that readily solidifies, m.p. 46–48°.

Yield = 3·5–4·5 g. (43–55%).

## Reviews

(a) "α-Pyrones", "Heterocyclic Compounds", ed. R. C. Elderfield, Vol. I, p. 354.
(b) "α-Pyrones", "Chemistry of Carbon Compounds", ed. E. H. Rodd, Vol. IV B, p. 815.

(c) "The Chemistry of Monocyclic $\alpha$- and $\gamma$-Pyrones", L. F. Cavalieri, *Chem. Rev.*, 1947, **41**, 525.
(d) "Polyphosphoric Acid as a Reagent in Organic Chemistry", F. D. Popp and W. E. McEwen, *Chem. Rev.*, 1958, **58**, 321.
(e) "Polyphosphoric Acid as a Reagent in Organic Chemistry", F. Uhlig and H. R. Snyder, *Adv. Org. Chem.*, 1960, **1**, 35.

## Reference

1. German Patent 1,159,962 (*Chem. Abs.*, 1964, **60**, 13227).

# 6 BENZODERIVATIVES OF SIX-MEMBERED SYSTEMS CONTAINING ONE HETEROATOM

## 3-METHYL-1-PHENYLISOQUINOLINE

In the Pictet–Gams modification[1] of the Bischler–Napieralski isoquinoline synthesis,[2] an *N*-acylated derivative of a β-hydroxy-β-phenylethylamine (e.g. IV) is cyclodehydrated to an isoquinoline.

In this example, 3-methyl-1-phenylisoquinoline (V) results from cyclization of 2-benzamido-1-hydroxy-1-phenylpropane (IV),[3] which is itself obtained by a 3-stage synthesis from propiophenone (I). Firstly, nitrosation with isopentyl nitrite yields isonitrosopropiophenone (II), which on hydrogenation in the presence of palladium–charcoal and hydrogen chloride affords 2-amino-1-hydroxy-1-phenylpropane hydrochloride (III).[4] Benzoylation of the latter yields 2-benzamido-1-hydroxy-1-phenylpropane (IV).[3]

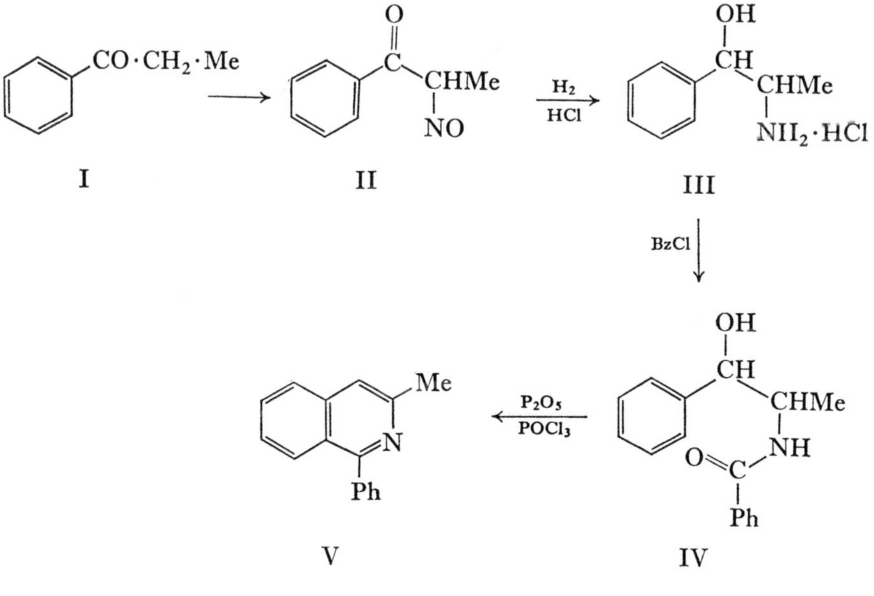

### Isonitrosopropiophenone (II)

In a 250-ml. 3-necked, round-bottomed flask fitted with gas-inlet tube, dropping-funnel and reflux condenser with drying tube, magnetically stir a solution of propiophenone (26·8 g.; 26·6 ml.) in ether (125 ml.). Pass a stream of dry hydrogen chloride gas (for method of generation, see p. 113) through the solution at the rate of ca. 2–3 bubbles/sec. and simultaneously add isopentyl nitrite (23·4 g.; 27 ml.) portionwise from the dropping-funnel over 0·5 hr. (After each addition, the solution takes on a deep yellow colour, which disappears as the reaction proceeds. The flask contents eventually boil gently.) When addition is complete, continue bubbling hydrogen chloride for 15 min., then allow the solution to stand overnight at room temperature.

Stir the ethereal solution into a mixture of 10% sodium hydroxide (250 ml.) and ice, then separate the two layers. Retain the aqueous layer, and extract the ethereal layer with 2N sodium hydroxide (4 × 50 ml.). Combine the extracts with the separated aqueous layer, and mechanically stir the solution in an ice–water bath while acidifying with concentrated hydrochloric acid. Filter off the crude product and wash with water.

Isonitrosopropiophenone crystallizes from benzene (ca. 3 ml./g.) as shining white needles, m.p. 113–115°.

Yield = 8·5–10 g. (26–31%).

### 2-Amino-1-hydroxy-1-phenylpropane hydrochloride (III)

To a solution of isonitrosopropiophenone (4·89 g.) in ethanolic hydrogen chloride (5% hydrogen chloride gas in dry ethanol) (75 ml.), add "5% palladium–charcoal catalyst" (1 g.), and shake the mixture in hydrogen at room temperature.[5]

**Note**

The reaction seldom starts immediately and may require an induction period of up to 1 hr.

When hydrogen uptake ceases (after 1·5–2 hr.), filter off the catalyst, and evaporate the filtrate to dryness under reduced pressure to obtain the crude product as a white crystalline solid.

2-Amino-1-hydroxy-1-phenylpropane hydrochloride crystallizes from ethanol (ca. 5 ml./g.) as fine white needles, m.p. 182–184°.

Yield = 2–2·5 g. (35–44%).

### 2-Benzamido-1-hydroxy-1-phenylpropane (IV)

In a 100-ml. conical flask, suspend 2-amino-1-hydroxy-1-phenylpropane hydrochloride (1·88 g.) in ether (20 ml.). Shake the suspension with 10% aqueous sodium hydroxide (8 ml.), and to the resulting mixture add benzoyl chloride (1·4 g.; 1·15 ml.). Carefully shake with cooling for 10 min., and evaporate off the solvent from the bulky precipitate to obtain the crude product as a white mass.

2-Benzamido-1-hydroxy-1-phenylpropane crystallizes from 50% aqueous ethanol (ca. 25 ml./g.) as shining white needles, m.p. 142–143°.

Yield = 2–2·2 g. (79–86%).

### 3-Methyl-1-phenylisoquinoline (V)

To a suspension of 2-benzamido-1-hydroxy-1-phenylpropane (2 g.) in dry xylene (50 ml.), add phosphorus pentoxide (20 g.) and phosphorus oxychloride (20 ml.), and heat the mixture under reflux for 2·5 hr. Cool the flask and contents, and then cautiously, with ice–water cooling, destroy excess of dehydrating agent by the gradual addition of water (100 ml.). Transfer the resulting two-phase system into a separating funnel and separate the layers. Retain the aqueous layer and extract the xylene layer with water (25 ml.). Unite the aqueous solutions, and after washing with benzene (25 ml.), make strongly alkaline with 30% sodium hydroxide, adding water if necessary to keep the inorganic material in solution. Extract the liberated isoquinoline with benzene (3 × 50 ml.) and after drying the combined extract ($MgSO_4$), reduce its volume to ca. 25 ml., and saturate with dry hydrogen chloride gas. Evaporate the resulting solution to dryness to obtain the crude isoquinoline hydrochloride as a light-brown viscous oil.

3-Methyl-1-phenylisoquinoline hydrochloride crystallizes (with difficulty) from a mixture of isopropanol and petroleum ether (b.p. 80–100°) as clumps of fine needles, m.p. 225° (decomp.).

To the crystalline hydrochloride, add saturated sodium hydrogen carbonate solution (25 ml.), and stir the resulting semi-solid until solidification is complete. Filter off the crude product, wash with water and dry.

3-Methyl-1-phenylisoquinoline crystallizes from benzene (ca. 10 ml./g.) (decolourizing carbon), as fine white needles, m.p. 126–127°.

Yield = 1–1·2 g. (58–70%).

## Spectra

(a) Infrared.[5]

## Reviews

(a) "Isoquinoline", "Heterocyclic Compounds", ed. R. C. Elderfield, Vol. 4, p. 344.
(b) "Isoquinoline and its Derivatives", "Chemistry of Carbon Compounds", ed. E. H. Rodd, Vol. IV A, p. 645.
(c) "The Chemistry of Isoquinolines", R. H. Manske, *Chem. Rev.*, 1942, **30**, 145.
(d) "The Preparation of 3,4-Dihydroisoquinolines and Related Compounds by the Bischler–Napieralski Reaction", *Org. Reactions*, 1951, **6**, 74.

## References

1. A. Pictet and A. Gams, *Ber.*, 1909, **42**, 2943; 1910, **43**, 2384.
2. A. Bischler and B. Napieralski, *Ber.*, 1893, **26**, 1903.
3. W. M. Whaley and W. H. Hartung, *J. Org. Chem.*, 1949, **14**, 650.
4. W. H. Hartung and J. C. Munch, *J. Amer. Chem. Soc.*, 1929, **51**, 2262.
5. M. Vajda and F. Ruff, *Acta Chim. Acad. Sci. Hung.*, 1964, **40**, 217.

# 6-CHLOROQUINOLINE

Primary aromatic amines with a vacant *ortho*-position, react with glycerol and concentrated sulphuric acid in the presence of an oxidizing agent to give quinolines (the Skraup reaction[1]). In the following example 6-chloroquinoline (I) is prepared from *p*-chloroaniline with *m*-nitrobenzenesulphonic acid[2] as the oxidizing agent.

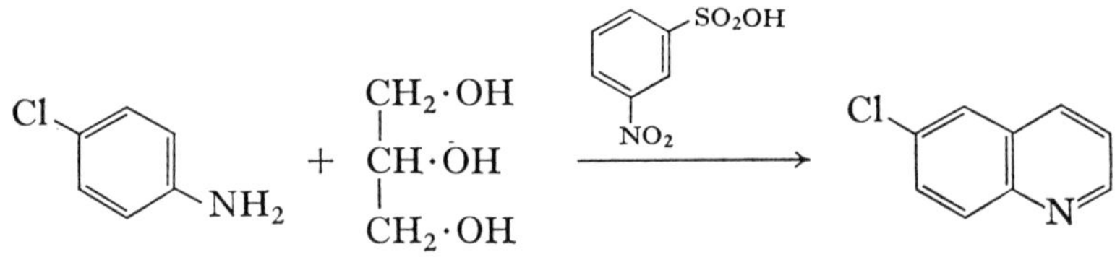


I

## 6-Chloroquinoline (I)

In a 500-ml. round-bottomed flask, prepare an intimate mixture of finely ground ferrous sulphate (1·75 g.), *p*-chloroaniline (7·6 g.), *m*-nitrobenzenesulphonic acid (5·8 g.), glycerol (18·7 g.; 14·8 ml.), and boric acid (3·1 g.). Add concentrated sulphuric acid (11 ml.), equip the flask with a reflux condenser and then cautiously heat the mixture in an oil-bath to 140° over a period of 15 min. Immediately the exothermic reaction commences remove the flask from the oil-bath, and allow the reaction to proceed for ca. 20 min. Replace the reaction flask in the oil-bath and continue heating at 130° for 2 hr. Cool, and pour the mixture into ice–water (300 ml.) contained in a 1-litre round-bottomed flask. Wash out the reaction vessel with water (200 ml.), and basify the combined aqueous solution with 30% sodium hydroxide. Steam-distil the mixture, and then extract the distillate with chloroform (2 × 60 ml.). Combine the extracts, dry ($MgSO_4$), and evaporate off the solvent to obtain the product as a pale-yellow oil that solidifies on standing.

6-Chloroquinoline crystallizes (with difficulty) from ether (ca. 4 ml./g.) as pale-yellow prisms, m.p. 40–41°.

Yield = 7–7·5 g. (72–77%).

### Spectra

Ultraviolet.[3]

### Reviews

(a) "The Skraup Synthesis of Quinolines", R. H. F. Manske and M. Kulka, *Org. Reactions*, 1953, **7**, 59.
(b) "Heterocyclic Nitrogen Compounds, Part IIa. Hexacyclic Compounds; Pyridine, Quinoline and Isoquinoline", F. W. Bergstrom, *Chem. Rev.*, 1944, **35**, 77.
(c) "The Chemistry of Quinoline", "Heterocyclic Compounds", ed. R. C. Elderfield, Vol. 4, p. 1.
(d) "Quinoline and its Derivatives", "Chemistry of Carbon Compounds", ed. E. H. Rodd, Vol. IV A, p. 584.

### References

1. H. Skraup, *Monatsh.*, 1880, **1**, 316.
2. W. P. Utermohlen jun., *J. Org. Chem.*, 1943, **8**, 544.
3. S. B. Knight, R. H. Wallick, and C. Balch, *J. Amer. Chem. Soc.*, 1955, **77**, 2577.

## 4-METHYL-2-PHENYLQUINOLINE

Aryl lithium compounds add across the azomethine bond of the quinoline nucleus to give products, that on heating lose the elements of lithium hydride to yield 2-phenylquinolines.[1] 4-Methylquinoline (lepidine) reacts with phenyllithium to give 4-methyl-2-phenylquinoline (I), together with a small amount of 1,2-dihydro-2,2-diphenyl-4-methylquinoline (II).[2]

$$Ph\cdot Br + Li \longrightarrow Ph\cdot Li$$

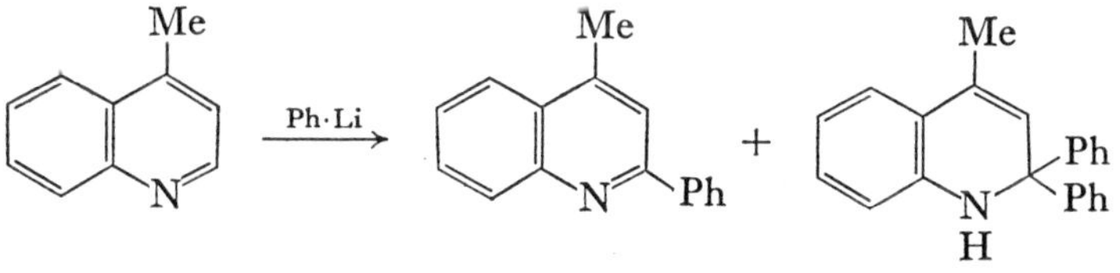


I II

### Phenyl-lithium

Place sodium-dried ether (40 ml.) into a 250-ml. 3-necked, round-bottomed flask fitted with reflux condenser, mercury-sealed or magnetic stirrer, and 100-ml. dropping-funnel, containing a solution of freshly distilled bromobenzene (15·7 g.;

10·5 ml.) in sodium-dried ether (60 ml.). Flush out the apparatus with dry (concentrated sulphuric acid) nitrogen gas, and equip the condenser with a drying tube ($CaCl_2$). To the ethereal solution, add small pieces of freshly cut lithium metal (1·45 g.), and then add a portion (10 ml.) of the ethereal bromobenzene solution. Stir vigorously until the reaction commences and the ether begins to reflux, then add the remainder of the bromobenzene solution at a rate such that the ether refluxes gently (ca. 15 min.). Continue to heat the solution under reflux with constant stirring, until all the lithium metal has dissolved (ca. 2 hr.), when the solution is ready for use in the next stage.

### 4-Methyl-2-phenylquinoline (I)

To the prepared ethereal solution of phenyl-lithium, add over a period of 5 min. a solution of freshly distilled 4-methylquinoline (14·3 g.; 13·1 ml.) in sodium-dried ether (30 ml.). Heat the mixture under reflux with stirring on a water-bath for 0·5 hr., then cautiously pour into a solution of 6N hydrochloric acid (60 ml.) and ice–water (200 ml.). Filter off the resulting precipitate of 4-methyl-2-phenylquinoline hydrochloride and 1,2-dihydro-2,2-diphenyl-4-methylquinoline, and add the mixture to 2N sodium hydroxide (25 ml.). Extract the liberated 4-methyl-2-phenylquinoline with ether (2 × 30 ml.), and filter off the insoluble by-product (II). Combine the extracts, dry ($MgSO_4$), and evaporate off the ether to obtain the crude product as a pale-yellow oil that solidifies on cooling.

4-Methyl-2-phenylquinoline crystallizes from petroleum ether (b.p. 60–80°; ca. 10 ml./g.) as pale-yellow plates, m.p. 64–65°.

Yield = 10–12 g. (45·5–55%).

The ether-insoluble 1,2-dihydro-2,2-diphenyl-4-methylquinoline crystallizes from ethanol (ca. 6 ml./g.) as colourless prisms, m.p. 163–164°.

Yield = 3–3·5 g.

**Reviews**

"Quinolines", see review references on p. 83.

**References**

1. K. Ziegler and H. Zeiser, *Annalen*, 1931, **485**, 188.
2. N. N. Goldberg and R. Levine, *J. Amer. Chem. Soc.*, 1955, **77**, 3647.

## 2,4,6-TRIMETHYLQUINOLINE

1,3-Dicarbonyl compounds condense with primary aromatic amines to give anils, which on treatment with concentrated sulphuric acid yield quinolines (the

Combes reaction[1]). Interaction of acetylacetone with *p*-toluidine gives the anil (I), which is cyclized by concentrated sulphuric acid to 2,4,6-trimethylquinoline (II).[2]

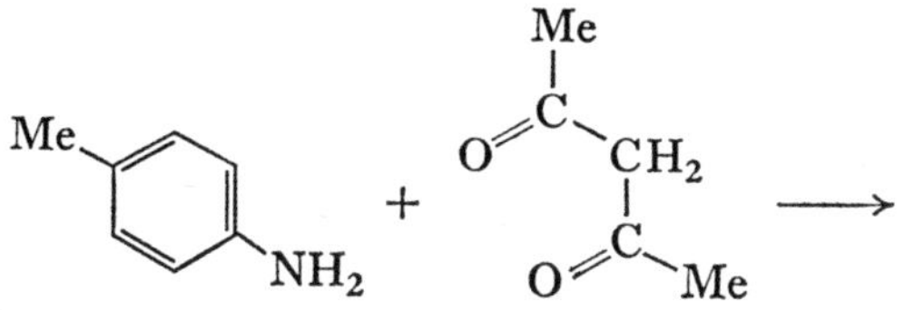

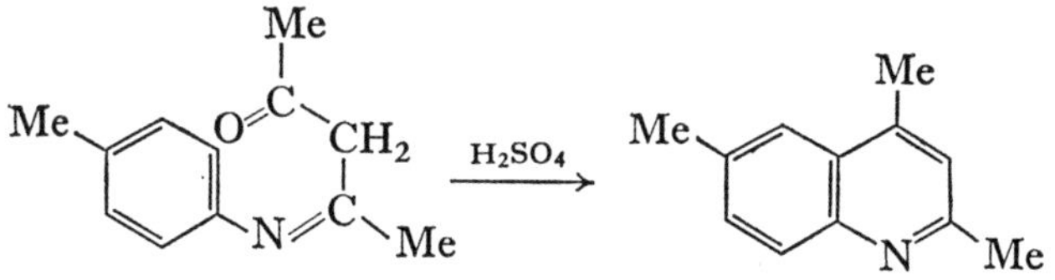

I II

### β-(*p*-Toluidino)propenyl methyl ketone (I)

In a 250-ml. round-bottomed flask fitted with a Dean and Stark apparatus[3] and reflux condenser, heat under reflux a solution of *p*-toluidine (10·7 g.) and acetylacetone (10 g.; 10·2 ml.) in xylene (50 ml.). When the theoretical quantity of water (2 ml.) has been collected (after ca. 2 hr.), distil off the excess of solvent, then cool and triturate the residue with cold petroleum ether (b.p. 40–60°; 30 ml.) until solidification occurs. Filter the mixture, and wash the residue with cold petroleum ether (b.p. 40–60°; 2 × 30 ml.).

*β*-(*p*-Toluidino)propenyl methyl ketone is obtained as pale-yellow plates, m.p. 65–66°, sufficiently pure for direct use in the next stage.

Yield = 7·5–8 g. (39·5–42·5%).

### 2,4,6-Trimethylquinoline (II)

To warm concentrated sulphuric acid (25 ml.) contained in a 250-ml. conical flask, add portionwise the prepared anil (I) (4 g.). When addition is complete, heat the mixture on a water-bath at 100° for 0·5 hr., then cool and cautiously pour into ice–water (200 ml.). Basify the resulting solution by adding solid sodium carbonate,† and filter off the precipitated product.

2,4,6-Trimethylquinoline dihydrate crystallizes (with difficulty) from 60% aqueous ethanol (ca. 8 ml./g.) as colourless needles, m.p. 62–63°.

Yield = 3–3·2 g. (68·5–73%).

**Spectra**

Proton magnetic resonance.[4]

† The sulphate salt of the quinoline may precipitate at this point.

## Reviews

"Quinolines", see review references on p. 83.

## References

1. A. Combes, *Bull. Soc. chim. France*, 1888, **49**, 89.
2. E. Roberts and E. E. Turner, *J. Chem. Soc.*, 1927, 1832.
3. A. I. Vogel, "A Textbook of Practical Organic Chemistry", Longmans, Green & Co., London, 3rd edn., 1956, p. 429.
4. W. Seiffert, *Angew. Chem.* (*Internat. edn.*), 1962, **1**, 215.

# 2-METHYL-4-QUINOLONE

β-Keto-esters condense with aromatic amines under mild conditions to yield anils, which cyclize on heating in an inert solvent to the corresponding 4-quinolones (the Conrad–Limpach synthesis[1]). Interaction of ethyl acetoacetate with aniline at low temperature (<100°) yields the anil (I)[2] that on heating in diphenyl ether at 200° cyclizes to give 2-methyl-4-quinolone (II).[3]

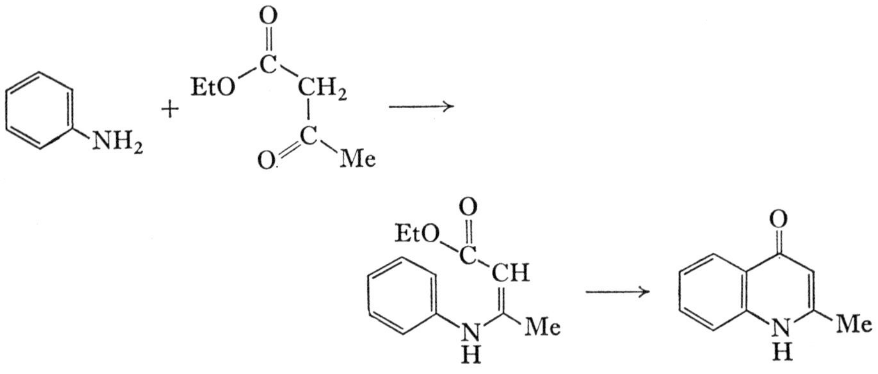


I II

### Ethyl β-anilinocrotonate (I)

In a 250-ml. round-bottomed flask, fitted with a Dean and Stark apparatus[4] and reflux condenser, heat under reflux a solution of freshly distilled aniline (9·3 g.; 9·1 ml.), ethyl acetoacetate (13 g.; 12·7 ml.) and glacial acetic acid (1 ml.) in benzene (100 ml.). When the theoretical quantity of water (2 ml.) has been collected (after ca. 3 hr.), evaporate off the solvent and transfer the crude product into a 25-ml. pear-shaped flask, and distil under water-pump vacuum.

Ethyl β-anilinocrotonate is obtained as a pale-yellow oil, b.p. 162–166°/12 mm. Yield = 14·5–15 g. (70·5–73%).

### 2-Methyl-4-quinolone (II)

To boiling diphenyl ether (25 ml.), contained in a 250-ml. 2-necked, round-bottomed flask equipped with reflux condenser and dropping-funnel, add dropwise over ca. 5 min. ethyl β-anilinocrotonate (5·1 g.). Heat the solution under reflux for 0·5 hr., then cool, and add petroleum ether (b.p. 60–80°; 100 ml.). Stir the mixture until a white solid is formed, then filter and wash the residue with petroleum ether (b.p. 60–80°; 4 × 50 ml.).

2-Methyl-4-quinolone crystallizes from water (ca. 20 ml./g.) as colourless prisms, m.p. 232–234°.

Yield = 3·1–3·5 g. (78–88%).

For a discussion on the tautomerism of 4-quinolones see ref. 5.

### Spectra

(a) Ultraviolet.[6]
(b) Infrared.[7]

### Reviews

(a) "The Chemistry of 4-Hydroxyquinolines", R. H. Reitsema, *Chem. Rev.*, 1948, **43**, 43.
(b) "Conrad–Limpach Reaction", "Heterocyclic Compounds", ed. R. C. Elderfield, Vol. 4, p. 30.
(c) "Quinolinols or Hydroxyquinolines", "Heterocyclic Compounds", ed. R. C. Elderfield, Vol. 4, p. 136.
(d) "Hydroxyquinolines", "Chemistry of Carbon Compounds", ed. E. H. Rodd, Vol. IV A, p. 590.

### References

1. M. Conrad and L. Limpach, *Ber.*, 1887, **20**, 944, 948.
2. G. A. Reynolds and C. R. Hauser, *Org. Synth.*, 1955, Coll. Vol. III, 374.
3. G. A. Reynolds and C. R. Hauser, *Org. Synth.*, 1955, Coll. Vol. III, 593.
4. A. I. Vogel, "A Textbook of Practical Organic Chemistry", Longmans, Green & Co., London, 3rd edn., 1956, p. 429.
5. A. R. Katritzky and J. M. Lagowski, *Adv. Heterocyclic Chem.*, 1963, **1**, 347.
6. J. K. Landquist, *J. Chem. Soc.*, 1951, 1038.
7. N. J. McCorkindale, *Tetrahedron*, 1961, **14**, 223.

## 4-METHYL-2-QUINOLONE

β-Ketoanilides cyclize under strong acid conditions to give substituted 2-quinolones (the Knorr synthesis[1]; see p. 88). Acetoacetanilide[2] on heating with concentrated sulphuric acid cyclizes to 4-methyl-2-quinolone (I).[3]

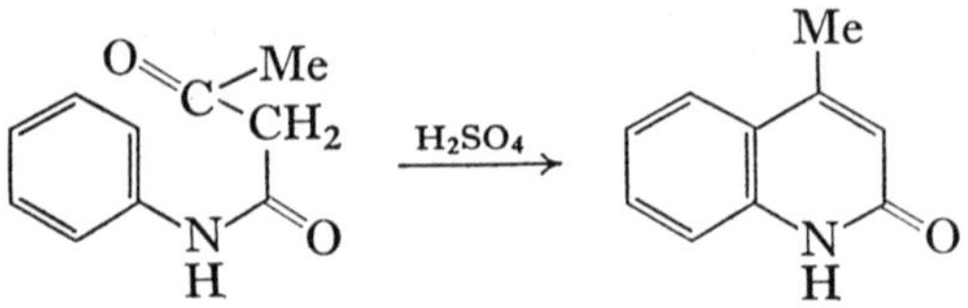


I

## 4-Methyl-2-quinolone (I)

Heat concentrated sulphuric acid (20 ml.) in a 100-ml. conical flask to 75° on a water-bath, and add portionwise acetoacetanilide (8·8 g.) at a rate such that the temperature of the reaction mixture remains at 70–75° (occasional cooling may be necessary). When addition is complete, heat the mixture on a water-bath at 100° for 1 hr., then cool and cautiously stir into ice–water (200 ml.). Filter off the product and dry at 100°.

4-Methyl-2-quinolone crystallizes from ethanol (ca. 12 ml./g.) as colourless plates, m.p. 220–222°.

Yield = 6–6·5 g. (75·5–81·5%).

For a discussion on the tautomerism of 2-quinolones see ref. 4.

### Spectra

(a) Ultraviolet.[5]
(b) Infrared.[5]

### Reviews

"Quinolones", see review references on p. 87.

### References


1. L. Knorr, *Annalen*, 1886, **236**, 83.
2. J. W. Williams and J. A. Krynitsky, *Org. Synth.*, 1955, Coll. Vol. III, 10.
3. W. M. Lauer and C. E. Kaslow, *Org. Synth.*, 1955, Coll. Vol. III, 580.
4. A. R. Katritzky and J. M. Lagowski, *Adv. Heterocyclic Chem.*, 1963, **1**, 347.
5. J. A. Gibson, W. Kynaston, and A. S. Lindsey, *J. Chem. Soc.*, 1955, 4340.


## 4-PHENYL-2-QUINOLONE

*β*-Keto-esters condense with primary aromatic amines at elevated temperatures to give *β*-ketoanilides, which on treatment with strong acid yield 2-quinolones (the Knorr synthesis[1]). Interaction of ethyl benzoylacetate with aniline under reflux gives the anilide (I), which on heating with polyphosphoric acid (PPA) yields 4-phenyl-2-quinolone (II).[2]

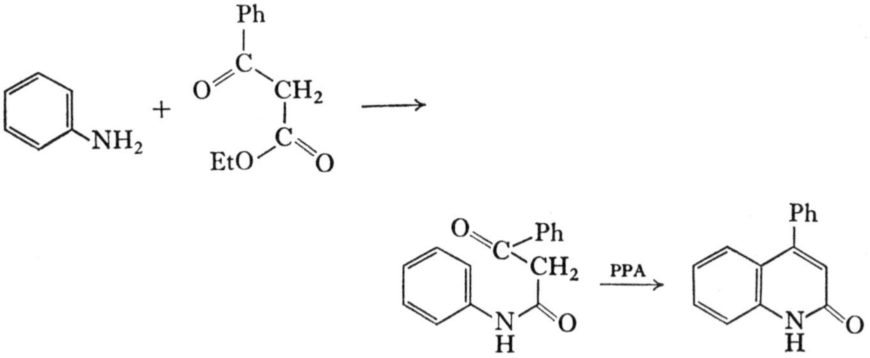

I II

### Benzoylacetanilide (I)

In a 250-ml. round-bottomed flask, heat a mixture of ethyl benzoylacetate (9·6 g.; 8·5 ml.) and aniline (4·65 g.; 4·55 ml.) under reflux for 1 hr. Cool the solution, add 10% sodium hydroxide (50 ml.), then heat on a water-bath at 100° for 5 min. Transfer the cooled reaction mixture into a separating funnel and extract with ether (2 × 50 ml.). Separate the aqueous layer, and precipitate the product by addition of excess of acetic acid. Filter the reaction mixture, and dry the residue at 80°.

Benzoylacetanilide crystallizes from petroleum ether (b.p. 80–100°; ca. 10 ml./g.) as colourless plates, m.p. 105–106°.

Yield = 7–8 g. (58·5–67%).

### 4-Phenyl-2-quinolone (II)

Add polyphosphoric acid (12 g.) to benzoylacetanilide (1 g.) and stir at 145° for 0·5 hr. (see apparatus on p. 38 and use tetrachloroethane as heating liquid). Cool the reaction mixture to ca. 80° and then stir into cold water (100 ml.). Filter the mixture, and wash the residue successively with 10% sodium hydroxide (50 ml.) and water (100 ml.).

4-Phenyl-2-quinolone crystallizes from 60% aqueous ethanol (ca. 12 ml./g.) (decolourizing carbon) as colourless needles, m.p. 257–259°.

Yield = 0·7–0·8 g. (75·5–86%).

**Reviews**

(a) "Knorr Synthesis of Quinolines", "Heterocyclic Compounds", ed. R. C. Elderfield, Vol. 4, p. 30.

(b) "Quinolones", see review references on p. 87.

(c) "Polyphosphoric Acid", see review references on p. 78.

**References**

1. L. Knorr, *Annalen*, 1886, **236**, 83.
2. B. Staskun and S. S. Israelstam, *J. Org. Chem.*, 1961, **21**, 3191.

## 9-ACRIDONE

*N*-Phenylanthranilic acid (I), prepared by condensing *o*-chlorobenzoic acid with aniline in the presence of a copper catalyst (the Ullmann reaction[1]), cyclizes on treatment with hot concentrated sulphuric acid to 9-acridone (II).

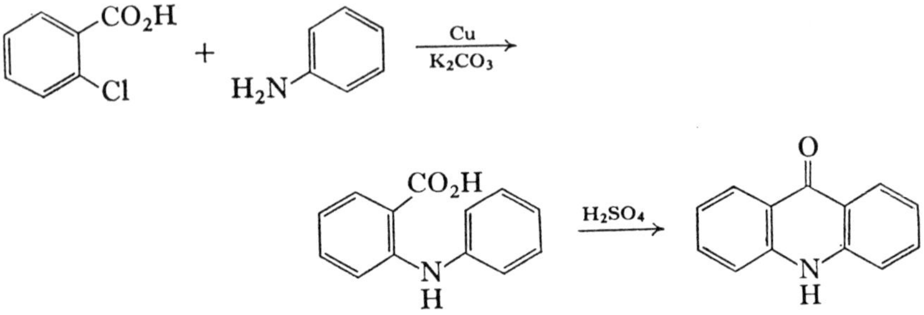


I II

### *N*-Phenylanthranilic acid (I)

In a 250-ml. round-bottomed flask heat under reflux a mixture of aniline (15·5 g.; 15·2 ml.), *o*-chlorobenzoic acid (4·1 g.), potassium carbonate (4·1 g.), and copper bronze powder (1 g.) for 2 hr. Cool, and to the semi-solid mass add a mixture of concentrated hydrochloric acid (25 ml.) and water (25 ml.). Shake the resulting mixture vigorously, then filter and dissolve the residue in 4N sodium hydroxide (20 ml.). Add decolourizing carbon, boil for 2 min., then filter the hot mixture. Cool, acidify the filtrate with concentrated hydrochloric acid, and filter off the precipitated product. Wash the residue with water (100 ml.) and dry at 100°.

*N*-Phenylanthranilic acid is obtained as a mauve solid, m.p. 184–185°, sufficiently pure for direct use in the next stage.

Yield=2·1–2·5 g. (37·5–45%).

### 9-Acridone (II)

In a 100-ml. conical flask, heat a solution of *N*-phenylanthranilic acid (2·5 g.) in concentrated sulphuric acid (10 ml.) on a water-bath at 100° for 3 hr. Cool the solution, and cautiously stir into ice–water (100 ml.). Boil the resulting mixture for 5 min., then cool and filter off the crystalline product.

9-Acridone crystallizes from glacial acetic acid (ca. 10 ml./g.) as yellow needles, m.p. 348–350°.

Yield = 1·4–1·7 g. (61–74·5%).

## Spectra

(a) Ultraviolet.[2]
(b) Infrared.[3]
(c) Proton magnetic resonance.[4]

## Reviews

(a) A. Albert, "The Acridines", Edward Arnold (publishers) Ltd., London, 1966.
(b) "Copper Promoted Reactions in Aromatic Chemistry", R. G. R. Bacon and H. A. O. Hill, *Quart. Rev.*, 1965, **19**, 95.
(c) "The Ullmann Synthesis of Biaryls, '1945–1963'", P. E. Fanta, *Chem. Rev.*, 1964, **64**, 613.
(d) "Acridines", "The Chemistry of Heterocyclic Compounds", ed. A. Weissberger, Vol. 9, p. 11.
(e) "Synthesis and Properties of Acridines", "Heterocyclic Compounds", ed. R. C. Elderfield, Vol. 4, p. 534.
(f) "Acridone and its Derivatives", "Chemistry of Carbon Compounds", ed. E. H. Rodd, Vol. IV A, p. 685.

## References

1. C. F. H. Allen and G. H. W. McKee, *Org. Synth.*, 1943, Coll. Vol. II, 15.
2. H. Kokubun, *Z. Elektrochem.*, 1958, **62**, 599.
3. J. R. Price and J. B. Willis, *Austral. J. Chem.*, 1959, **12**, 589.
4. J. P. Kokko and J. H. Goldstein, *Spectrochim. Acta*, 1963, **19**, 1119.

# PHENANTHRIDINE

Phenanthridines are obtained by cyclization of *o*-acylaminobiphenyls with polyphosphoric acid (PPA).[1] *o*-Formamidobiphenyl (I) cyclizes to give an excellent yield of phenanthridine (II).

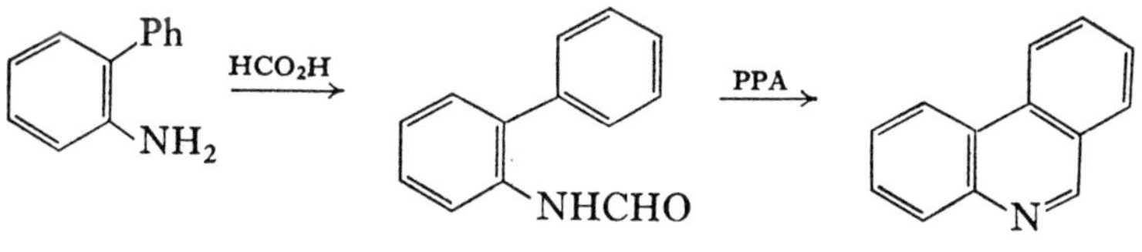


I II

### *o*-Formamidobiphenyl (I)

Heat a solution of *o*-aminobiphenyl[2] (2 g.) in 98% formic acid (15 ml.) under reflux for 1 hr. Cool, pour the solution into cold water (50 ml.), and stir until the product solidifies. Filter the mixture and dry the residue at 100°.

*o*-Formamidobiphenyl crystallizes from aqueous ethanol (ca. 5 ml./g.) as colourless needles, m.p. 74–75°.

Yield = 1·9–2·1 g. (81·5–90%).

### Phenanthridine (II)

To polyphosphoric acid (25 g.), add *o*-formamidobiphenyl (1·8 g.) and stir at 155° for 1 hr. (see apparatus on p. 38 and use cyclohexanone as heating liquid). Cool the mixture to ca. 80°, and stir into cold water (100 ml.). Basify the aqueous solution with concentrated ammonium hydroxide, then cool and filter off the product and dry at 80°.

Phenanthridine crystallizes from petroleum ether (b.p. 80–100°; ca. 8 ml./g.) as colourless needles, m.p. 102–103°.

Yield = 1·4–1·5 g. (85·5–91·5%).

## Spectra

(a) Ultraviolet.[3]
(b) Proton magnetic resonance.[4]

## Reviews

(a) "The Chemistry of Phenanthridine and its Derivatives", R. S. Theobald and K. Schofield, *Chem. Rev.*, 1950, **46**, 171.
(b) "Comparison of Phenanthridine with other Aza-aromatic Heterocycles", J. Eisch and H. Gilman, *Chem. Rev.*, 1957, **57**, 525.
(c) "4- and 5-Azaphenanthrenes", "The Chemistry of Heterocyclic Compounds", ed. A. Weissberger, Vol. 12, p. 271.
(d) "Phenanthridine", "Heterocyclic Compounds", ed. R. C. Elderfield, Vol. 4, p. 564.
(e) "Phenanthridine and its Derivatives", "Chemistry of Carbon Compounds", ed. E. H. Rodd, Vol. IV A, p. 691.
(f) "Polyphosphoric Acid", see review references on p. 78.

## References

1. E. C. Taylor, jun. and N. W. Kalenda, *J. Amer. Chem. Soc.*, 1954, **76**, 1699.
2. See preparation on p. 7.
3. G. M. Badger, R. S. Pearce, and R. Pettit, *J. Chem. Soc.*, 1951, 3199.
4. E. V. Donckt, R. H. Martin, and F. Geerts-Evrard, *Tetrahedron*, 1964, **20**, 1495.

## 6-PHENANTHRIDONE

Beckmann rearrangement[1] of alicyclic ketoximes yields cyclic secondary amides. Treatment of fluorenone oxime (I)[2] with polyphosphoric acid (PPA) gives an excellent yield of 6-phenanthridone (II).[3]

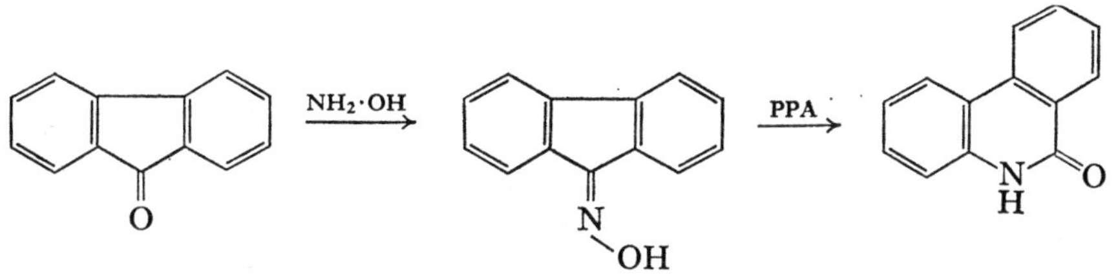


I II

### Fluorenone oxime (I)

To a solution of fluorenone[4] (8·5 g.) in ethanol (70 ml.), contained in a 250-ml. conical flask, add a solution of hydroxylamine hydrochloride (6·6 g.) in water (30 ml.). Heat the resulting mixture on a water-bath at 100° for 0·5 hr., then add sodium acetate (12 g.) and heat for a further 1 hr. Cool the mixture, filter off the product, and dry at 100°.

Fluorenone oxime is obtained as a pale-yellow solid, m.p. 195–197°, sufficiently pure for direct use in the next stage.

Yield = 8·5–9 g. (92·5–97·5%).

### 6-Phenanthridone (II)

To polyphosphoric acid (60 g.) add fluorenone oxime (2 g.), and stir at 175° for 15 min. (see apparatus on p. 38 and use *p*-cymene as heating liquid). Cool the mixture to ca. 80° and then stir into cold water (300 ml.). Filter off the product, wash with cold water (3 × 50 ml.) and dry at 100°.

6-Phenanthridone crystallizes from ethanol (ca. 40 ml./g.) as colourless needles, m.p. 292–293°.

Yield = 1·6–1·8 g. (80–90%).

**Spectra**

Ultraviolet.[5]

**Reviews**

(a) "The Beckmann Rearrangement", L. G. Donaruma and W. Z. Heldt, *Org. Reactions*, 1960, **11**, 1.

(b) "Molecular Rearrangements", ed. P. de Mayo, Interscience Publishers Inc., New York, 1963, (part 1), p. 457. (P. A. S. Smith).
(c) "Phenanthridone and its Derivatives", "Heterocyclic Compounds", ed. R. C. Elderfield, Vol. 4, p. 604.
(d) "Hydroxyphenanthridines; Phenanthridones", "Chemistry of Carbon Compounds", ed. E. H. Rodd, Vol. IV A, p. 698.
(e) "Polyphosphoric Acid", see review references on p. 78.

## References

1. E. Beckmann, *Ber.*, 1886, **19**, 988.
2. F. J. Moore and E. H. Huntress, *J. Amer. Chem. Soc.*, 1927, **49**, 2618.
3. E. C. Horning, V. L. Stromberg, and H. A. Lloyd, *J. Amer. Chem. Soc.*, 1952, **74**, 5153.
4. D. A. Shirley, "Preparation of Organic Intermediates", J. Wiley and Sons, Inc., New York, 1951, p. 163.
5. C. B. Reese, *J. Chem. Soc.*, 1958, 895.

# CHROMONE-2-CARBOXYLIC ACID

On fusion with aluminium chloride, phenyl acetate undergoes a Fries rearrangement[1] yielding *o*- and *p*-hydroxyacetophenone, which are separated by steam-distillation. Chromone-2-carboxylic acid (II) results from Claisen acylation[2] of *o*-hydroxyacetophenone using diethyl oxalate, followed by cyclization and hydrolysis of the diketo-ester (I).

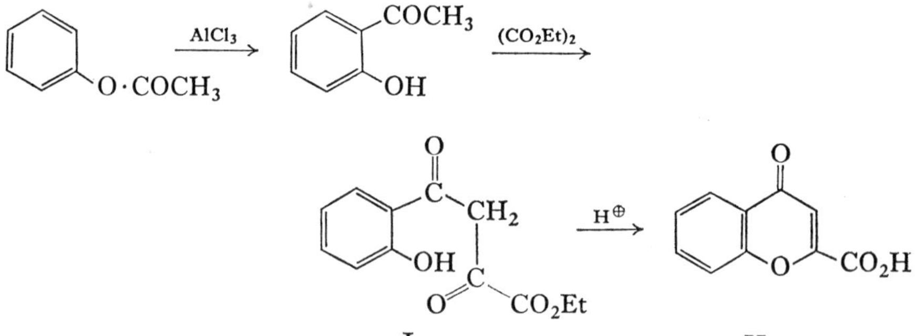


I II

## *o*-Hydroxyacetophenone

In a 250-ml. round-bottomed flask equipped with calcium chloride tube, heat a mixture of phenyl acetate (13·6 g.; 12·6 ml.) and anhydrous aluminium chloride (14·7 g.) in an oil-bath (temperature 165–170°) for 0·5 hr. Cool, and after grinding up the complex using a pestel and mortar, decompose the resulting powder by gradual addition to a mixture of crushed ice (250 g.) and concentrated hydrochloric acid (25 ml.).

Steam-distil the resulting oily mixture until all the steam-volatile *o*-hydroxyacetophenone has been removed (ca. 3 hr.), then extract the steam-distillate with ether (3 × 100 ml.). Combine the extracts, dry ($MgSO_4$), and after evaporating off the ether, distil the residual oil.

*o*-Hydroxyacetophenone is obtained as a pale-yellow oil, b.p. 200–210°.

Yield = 4–6 g. (29–44%).

### Chromone-2-carboxylic acid (II)

To a solution of sodium (1·38 g.) in ethanol (40 ml.) in a 100-ml. round-bottomed flask, add diethyl oxalate (3 ml.) and *o*-hydroxyacetophenone (2·72 g.), and heat the mixture under reflux for 0·5 hr. Cool, and fully precipitate the sodium salt of the diketone by addition of the yellow mass to ether (150 ml.). Allow the suspension to stand for 0·5 hr., then filter off the sodium salt and add this to 2N hydrochloric acid (50 ml.). Extract the liberated semi-solid with ether (3 × 20 ml.), combine the extracts and evaporate off the ether. Heat the residual diketone under reflux with a mixture of glacial acetic acid (10 ml.) and concentrated hydrochloric acid (10 ml.) for 0·5 hr. After cooling, filter off the crude product.

Chromone-2-carboxylic acid crystallizes from glacial acetic acid (ca. 10 ml./g.) as pale-yellow needles, m.p. 250–251° (decomp.).

Yield = 2·4–2·8 g. (63–74%).

### Spectra

Ultraviolet.[2]

### Reviews

(a) "Chromones", see review references on p. 100.
(b) "The Fries Reaction", A. H. Blatt, *Chem. Rev.*, 1940, **27**, 413.
(c) "The Fries Reaction", A. H. Blatt, *Org. Reactions*, 1942, **1**, 342.
(d) "The Acylation of Ketones to form β-Diketones or β-Ketoaldehydes", C. R. Hauser, F. W. Swamer, and J. T. Adams, *Org. Reactions*, 1954, **8**, 59.

### References

1. K. Fries and G. Finck, *Ber.*, 1908, **41**, 4271.
2. J. Schmutz, H. Lauener, R. Hirt, and M. Sanz, *Helv. Chim. Acta*, 1951, **34**, 767.

## 2-METHYL-5,6,7,8-TETRAHYDROCHROMONE

Enamines (α,β-unsaturated amines) (e.g. I) undergo alkylation and acylation at the β-position. Certain enamines are acylated by diketen yielding derivatives of γ-pyrone.[1] Interaction of 1-morpholinocyclohexene (I) (p. 49) and diketen leads to 2-methyl-5,6,7,8-tetrahydrochromone (III) via the morpholino intermediate (II).[1]

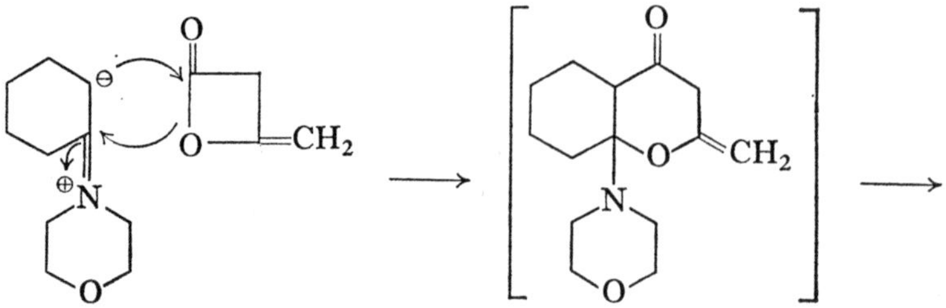


I II

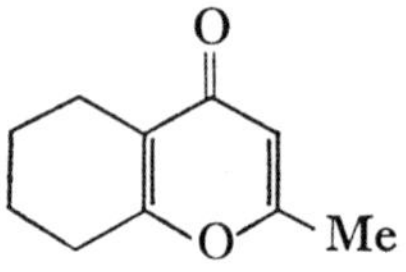


III

### 2-Methyl-5,6,7,8-tetrahydrochromone (III)

In a 250-ml. round-bottomed, 3-necked flask fitted with internal thermometer, drying tube and dropping-funnel, mechanically stir 1-morpholinocyclohexene (16·7 g.) at 60°. Add diketen (18 g.; 16·5 ml.) dropwise over 45 min. keeping the temperature at 60–65°, using cooling when necessary.

**Note**

If the addition is carried out at temperatures below 60°, the rate of reaction is slowed down considerably. This can lead to a build-up of unreacted materials with subsequent loss of control of the reaction.

When addition is complete, continue stirring, and heat at 70° for 2 hr., then allow to stand overnight at room temperature.

Fractionate the resulting reddish oil (which may show signs of crystallization) using a short column, and isolate the main fraction, approx. b.p. 115–140°/1 mm. (Discard a short fore-run and a high-boiling residue.) Dissolve the pale-yellow main fraction in trichloroethylene (60 ml.), and remove an alkali-soluble by-product† by washing with 2N sodium hydroxide (saturated with sodium chloride) (3 × 50 ml.). Combine the extracts, dry ($MgSO_4$), and evaporate off the trichloroethylene to obtain the crude product as a mass of yellow crystals.

2-Methyl-5,6,7,8-tetrahydrochromone crystallizes from cyclohexane (ca. 3 ml./g.) as pale-yellow prisms, m.p. 94–95°.

Yield = 6·5–10 g. (40–61%).

† The by-product is *N*-acetoacetylmorpholine.

## Reviews

(a) "Chromones", see review references on p. 100.
(b) "Enamines", J. Szmuszkovicz, *Adv. Org. Chem.*, 1963, **4**, 1.
(c) "Keten in Organic Synthesis", R. N. Lacey, *Adv. Org. Chem.*, 1960, **2**, 213.

## Reference

1. S. Hünig, E. Benzing, and K. Hübner, *Chem. Ber.*, 1961, **94**, 486.

# 7-HYDROXY-4-METHYLCOUMARIN (4-METHYLUMBELLIFERONE)

Under acidic conditions, phenols condense with ethyl acetoacetate, yielding coumarins (the von Pechmann reaction[1]). Resorcinol and ethyl acetoacetate in the presence of sulphuric acid yield 7-hydroxy-4-methylcoumarin (I).[2]

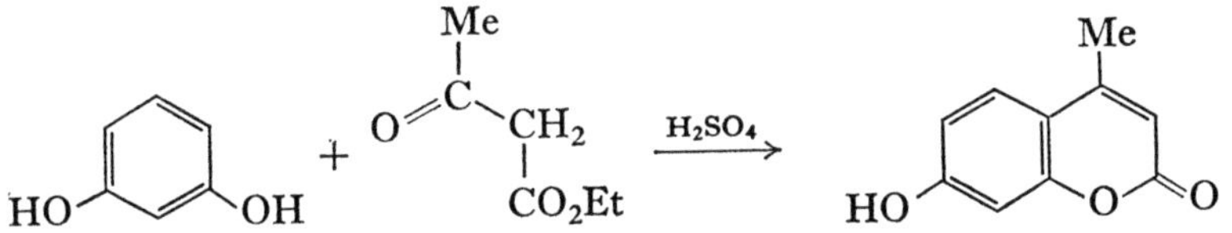


I

### 7-Hydroxy-4-methylcoumarin (I)

Heat a mixture of resorcinol (5·5 g.) and ethyl acetoacetate (6·5 g.; 6·35 ml.) with 75% sulphuric acid (50 ml.) in a 100-ml. conical flask on the water-bath at 100° for 0·5 hr. Cool the resulting dark-green solution, and stir into crushed ice (250 g.). Filter off the crude product, and wash with water (100 ml.).

7-Hydroxy-4-methylcoumarin crystallizes from methanol (ca. 8 ml./g.)† as pale-yellow plates, m.p. 185–186°.

Yield = 6–6·5 g. (68–74%).

## Spectra

(a) Ultraviolet.[3]
(b) Infrared.[4]
(c) Proton magnetic resonance.[5]

† This solubility refers to the monohydrate. If the crude product is dried at 100°, the anhydrous coumarin is produced which is virtually insoluble in alcohol, but which crystallizes (with difficulty) from benzene.

**Reviews**

(a) "Coumarins", "Heterocyclic Compounds", ed. R. C. Elderfield, Vol. 2, p. 173.
(b) "Coumarin and its Derivatives", "Chemistry of Carbon Compounds", ed. E. H. Rodd, Vol. IV B, p. 869.
(c) "The Chemistry of Coumarins", S. M. Sethna and N. M. Shan, *Chem. Rev.*, 1945, **36**, 1.
(d) "The Pechmann Reaction", S. Sethna and R. Phadke, *Org. Reactions*, 1953, **7**, 1.

**References**

1. H. von Pechmann and C. Duisberg, *Ber.*, 1883, **16**, 2119.
2. A. A. Shamshurin, *Chem. Abs.*, 1941, **35**, 3994.
3. B. N. Mattoo, *Trans. Faraday Soc.*, 1956, **52**, 1184.
4. M. E. Perel'son, *Zhur. obshchei Khim.*, 1963, **33**, 952.
5. S. S. Dharmatti, G. Govil, C. R. Kanekar, C. L. Khetrapal, and Y. P. Virmani, *Proc. Indian Acad. Sci.*, 1962, **56**, 71.


## BENZO[*f*]CHROMAN-4-ONE

Phenols yield cyanoethyl ethers when treated with acrylonitrile in the presence of base. Hydrolysis of the cyanoethyl ether followed by cyclization of the resulting acid yields a chroman-4-one. The cyanoethyl ether (I) of $\beta$-naphthol when treated with 85% sulphuric acid yields benzo[*f*]chroman-4-one (II) directly.[1]

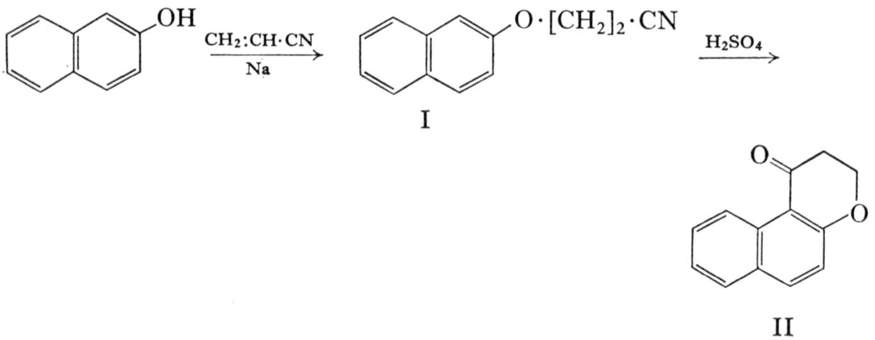


I

II

### β-Naphthyl 2-cyanoethyl ether (I)

To $\beta$-naphthol (24 g.) and acrylonitrile (16 g.; 20 ml.) in a 100-ml. round-bottomed flask, add sodium (0·4 g.) and heat the mixture under reflux for 3 hr.† Cool the resulting dark-brown liquid, and stir into a mixture of water (200 ml.) and concentrated hydrochloric acid (10 ml.), when a brownish oil is liberated which readily solidifies. Transfer the mixture into a separating funnel, and extract

† A yield of 79% is obtainable after much longer periods of heating.

with benzene (3 × 40 ml.). Combine the extracts, and wash the benzene solution successively with 2N sodium hydroxide (3 × 25 ml.) and water (25 ml.). Dry ($MgSO_4$), and evaporate off the benzene to obtain the crude product as a light-brown oil that readily solidifies.

β-Naphthyl 2-cyanoethyl ether crystallizes from ethanol (ca. 3 ml./g.) as lustrous white plates, m.p. 107–109°.

Yield = 10–12 g. (30–37%).

## Benzo[*f*]chroman-4-one (II)

In a 250-ml. conical flask, mechanically stir a solution of β-naphthyl 2-cyano-ethyl ether (10 g.) in 85% sulphuric acid (100 ml.) at room temperature for 2 hr. Stir the resulting deep-yellow solution into crushed ice (400 g.), and allow the mixture to stand in the refrigerator overnight. Extract the precipitated solid with ether (3 × 50 ml.) and wash the combined ethereal extracts successively with water, 2N sodium hydroxide (25 ml.) and water. Dry ($MgSO_4$), and after evaporating off the ether, distil the residue† (ca. 7 g.) under water-pump vacuum.

Benzo[*f*]chroman-4-one is obtained as an almost colourless oil, b.p. 207–208°/12 mm., that readily solidifies, m.p. 47–49°.

Yield = 6–6·5 g. (59–64%).

### Reviews

(a) "4-Chromanones", "Heterocyclic Compounds", ed. R. C. Elderfield, Vol. 2, p. 346.
(b) "Chromanones", "Chemistry of Carbon Compounds", ed. E. H. Rodd, Vol. IV B, p. 942.
(c) "Cyanoethylation", H. A. Bruson, *Org. Reactions*, 1949, **5**, 79.
(d) "Cyanoethylation, Decyanoethylation and Transcyanoethylation", P. F. Butskus, *Russ. Chem. Rev.*, 1961, **30**, 583.

### Reference

1. G. B. Bachman and H. A. Levine, *J. Amer. Chem. Soc.*, 1947, **69**, 2341.

## BENZO[*f*]CHROMONE

Dehydrogenation of chroman-4-ones is effected by triphenylmethyl per-chlorate, and involves the isolation of the intermediate 4-hydroxybenzopyrilium salt.[1]

† A slightly reduced yield may be obtained by crystallization of this residue from a benzene–petroleum ether (b.p. 80–100°) mixture as colourless needles, m.p. 50–51°.

Treatment of benzo[*f*]chroman-4-one (I) (p. 98) with triphenylmethyl perchlorate (obtained from interaction of triphenylcarbinol and perchloric acid)[2] yields the naphthopyrilium salt (II), which is converted by sodium bicarbonate into benzo[*f*]chromone (III).[3]

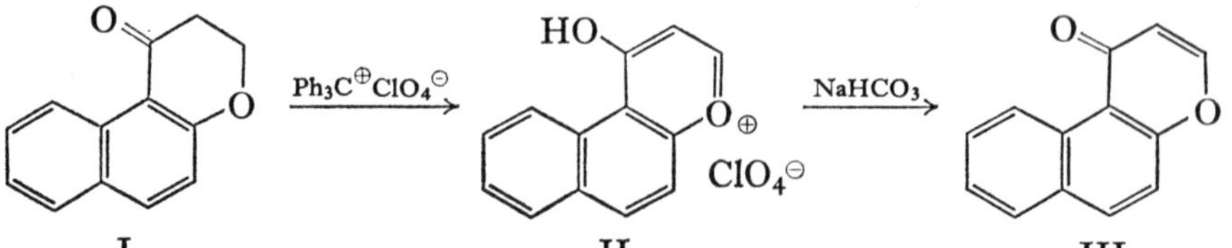


I II III

## Triphenylmethyl perchlorate

To a well-cooled, magnetically stirred solution of triphenyl carbinol (2 g.) in acetic anhydride (23 ml.), contained in a 100-ml. conical flask, add dropwise 70% perchloric acid (1·5 ml.). Continue stirring the mixture in the cooling bath for 1 hr., then filter off the product and wash well with ether. Dry in a vacuum desiccator.

Triphenylmethyl perchlorate is obtained as red prisms, m.p. 142–143°, sufficiently pure for direct use in the next stage.

Yield = 2·1–2·3 g. (79–87%).

## Benzo[*f*]chromone (III)

To a solution of benzo[*f*]chroman-4-one (0·99 g.) in glacial acetic acid (8 ml.) and acetic anhydride (1 ml.), in a 50-ml. round-bottomed flask, add freshly prepared triphenylmethyl perchlorate (1·71 g.). Heat the suspension on a water-bath at 100° for 0·5 hr., then evaporate the mixture to dryness under reduced pressure, and grind up the resulting yellow–brown solid with ether to remove triphenylmethane. Stir the residual naphthopyrilium salt thoroughly with saturated aqueous sodium hydrogen carbonate (20 ml.), and filter off the crude product.

Benzo[*f*]chromone crystallizes from petroleum ether (b.p. 60–80°; ca. 30 ml./g.) as pale-yellow elongated prisms, m.p. 99–101°.

Yield = 0·5–0·7 g. (51–71%).

### Reviews

(a) "Chromones", "Heterocyclic Compounds", ed. R. C. Elderfield, Vol. 2, p. 229.
(b) "Chromones", "Chemistry of Carbon Compounds", ed. E. H. Rodd, Vol. IV B, p. 888.
(c) "Benzopyrilium salts", "Chemistry of Carbon Compounds", ed. E. H. Rodd, Vol. IV B, p. 847.

**References**

1. A. Schönberg and G. Schütz, *Chem. Ber.*, 1960, **93**, 1466.
2. H. J. Dauben jun., L. R. Honnen, and K. M. Harmon, *J. Org. Chem.*, 1960, **25**, 1442.
3. A. O. Fitton, University of Salford, unpublished results.

## XANTHONE

Certain *o*-carboxyaryl fluoroborates react with phenols to produce xanthones.[1] Diazotization of anthranilic acid (I) yields *o*-carboxybenzenediazonium chloride, which is converted by fluoroboric acid into *o*-carboxybenzenediazonium fluoroborate (II).[2] The latter reacts with phenol to produce xanthone (III).[1]

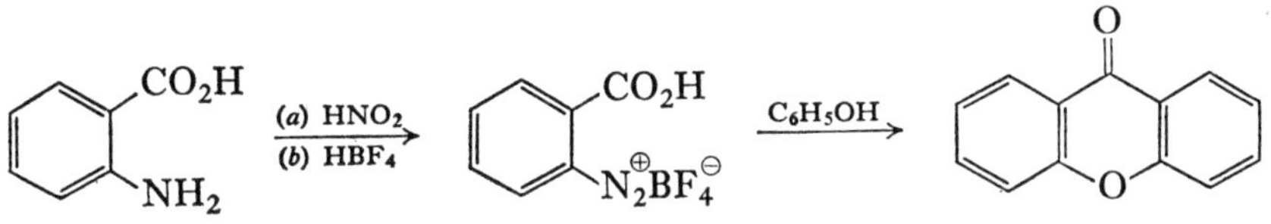

I II III

### *o*-Carboxybenzenediazonium fluoroborate (II)

In a 250-ml. conical flask, fitted with magnetic stirrer and internal thermometer, dissolve anthranilic acid (15 g.) in warm water (50 ml.) and add 3N hydrochloric acid (60 ml.). Cool the solution to 0° in an ice–salt bath, and with stirring add a solution of sodium nitrite (7·5 g.) in water (10 ml.), maintaining the temperature at 0°. After stirring for 5 min., add 40% fluoroboric acid (24 g.) and stir for a further 10 min. to complete the reaction. Filter off the precipitated product, wash successively with a little cold water and ether, then dry in a vacuum desiccator.

*o*-Carboxybenzenediazonium fluoroborate is obtained as light-brown plates, m.p. 125° (decomp.), sufficiently pure for direct use in the next stage.

Yield = 9–11 g. (35–42%).

### Xanthone (III)

To a magnetically stirred melt of phenol (5 g.) at ca. 130°, contained in a 50-ml. conical flask, add *o*-carboxybenzenediazonium fluoroborate (7·54 g.) portionwise over 10 min. Allow the resulting dark oil to cool to ca. 90°, then add this to hot 4N sodium hydroxide (150 ml.). After cooling to room temperature, filter off the crude product and wash with water.

Xanthone crystallizes from ethanol (ca. 15 ml./g.) as white needles, m.p. 173–174°.

Yield = 0·9–1·25 g. (14–20%).

## Spectra

(a) Ultraviolet.[3]
(b) Infrared.[4]
(c) Proton magnetic resonance.[5]

## Reviews

(a) "Xanthones", "Heterocyclic Compounds", ed. R. C. Elderfield, Vol. 2, p. 421.
(b) "Xanthones", "Chemistry of Carbon Compounds", ed. E. H. Rodd, Vol. IV B, p. 976.

## References

1. C. Sellers and H. Suschitzky, University of Salford, unpublished results.
2. J. F. Dippy and F. R. Williams, *J. Chem. Soc.*, 1934, 1466.
3. S. Kruger and F. G. Mann, *J. Chem. Soc.*, 1955, 2755.
4. M. St. C. Flett, *Spectrochim. Acta*, 1962, **18**, 1537.
5. C. T. Mathis and J. H. Goldstein, *Spectrochim. Acta*, 1964, **20**, 871.

# 7 SIX-MEMBERED SYSTEMS CONTAINING MORE THAN ONE HETEROATOM

## 4-CYANO-2,3-DIHYDRO-5,6-DIPHENYL-3-OXOPYRIDAZINE

Under the influence of base, the monohydrazones of certain α-dicarbonyl compounds condense with ethyl acetoacetate, diethyl malonate or ethyl cyanoacetate yielding 3-oxopyridazines.[1]

Benzil, hydrazine, and ethyl cyanoacetate in the presence of sodium ethoxide, condense directly to give 4-cyano-2,3-dihydro-5,6-diphenyl-3-oxopyridazine (I).[1]

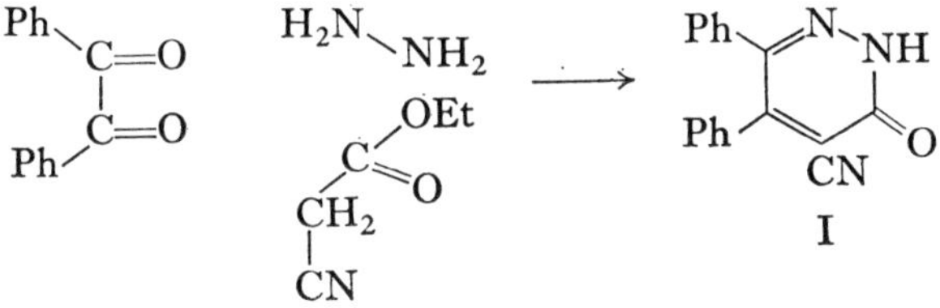

I

### 4-Cyano-2,3-Dihydro-5,6-diphenyl-3-oxopyridazine (I)

In a 250-ml. round-bottomed flask fitted with magnetic stirrer, dissolve ethyl cyanoacetate (5·65 ml.), benzil (10·5 g.), and hydrazine hydrate (2·5 g.) in ethanol (75 ml.), and to the mixture add a solution of sodium (1·15 g.) in ethanol (50 ml.). Stir and heat the resulting deep-red solution under reflux for 2 hr., then evaporate to dryness under reduced pressure. To the residual semi-solid, add water (100 ml.) and grind up to form a pale-yellow suspension. With stirring, add 2N hydrochloric acid until pH 6–7 is reached, then filter off the precipitated crude product.

4-Cyano-2,3-dihydro-5,6-diphenyl-3-oxopyridazine crystallizes from ethanol (ca. 10 ml./g.) (decolourizing carbon), as pale-yellow hexagonal plates, m.p. 270–272°.

Yield = 4·5–6 g. (33–44%).

For a discussion on the tautomerism of pyridazones, see ref. 2.

### Reviews

(a) "New Methods in the Chemistry of Pyridazones", K. Dury, *Angew. Chem.* (*Internat. edn.*), 1965, **4**, 292.
(b) "Pyridazones", "Heterocyclic Compounds", ed. R. C. Elderfield, Vol. 6, p. 113.
(c) "Pyridazine and its Substitution Products", "Chemistry of Carbon Compounds", ed. E. H. Rodd, Vol. IV B, p. 1201.

### References

1. P. Schmidt and J. Druey, *Helv. Chim. Acta*, 1954, **37**, 134.
2. A. R. Katritzky and J. M. Lagowski, *Adv. Heterocyclic Chem.*, 1963, **1**, 364.

## 2,4-DIOXO-6-METHYL-1,2,3,4-TETRAHYDROPYRIMIDINE (6-METHYLURACIL)

This general pyrimidine synthesis involves the condensation of urea or a urea derivative with a β-dicarbonyl compound.[1] In this example, thiourea and ethyl acetoacetate yield 6-methyl-4-oxo-1,2,3,4-tetrahydro-2-thiopyrimidine (I),[2] which is subsequently desulphurized[3] by chloroacetic acid to 2,4-dioxo-6-methyl-1,2,3,4-tetrahydropyrimidine (II).

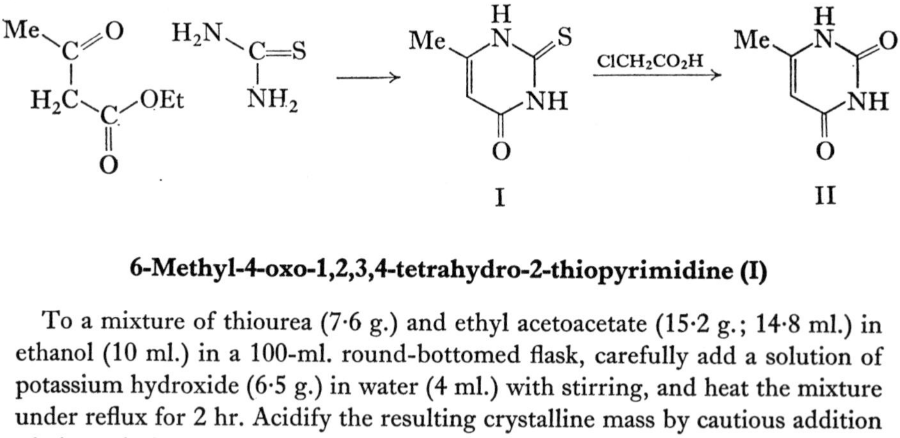


### 6-Methyl-4-oxo-1,2,3,4-tetrahydro-2-thiopyrimidine (I)

To a mixture of thiourea (7·6 g.) and ethyl acetoacetate (15·2 g.; 14·8 ml.) in ethanol (10 ml.) in a 100-ml. round-bottomed flask, carefully add a solution of potassium hydroxide (6·5 g.) in water (4 ml.) with stirring, and heat the mixture under reflux for 2 hr. Acidify the resulting crystalline mass by cautious addition of a hot solution of concentrated hydrochloric acid (20 ml.) in water (10 ml.), then cool and filter off the product.

6-Methyl-4-oxo-1,2,3,4-tetrahydro-2-thiopyrimidine is obtained as white crystals, m.p. >290° (decomp.), sufficiently pure for direct use in the next stage.

Yield = 10–12 g. (70–85%).

### 2,4-Dioxo-6-methyl-1,2,3,4-tetrahydropyrimidine (II)

Add the 6-methyl-4-oxo-1,2,3,4-tetrahydro-2-thiopyrimidine from the above experiment to a solution of chloroacetic acid (14 g.) in water (200 ml.) in a 500-ml. round-bottomed flask, and heat the mixture under reflux for 3 hr. in the fume cupboard. Filter the hot solution and allow the product to crystallize.

2,4-Dioxo-6-methyl-1,2,3,4-tetrahydropyrimidine is obtained as colourless prisms, m.p. >300°.

Yield=7–8 g. (55·5–63·5%, based on thiourea).

For a discussion on the tautomerism of oxopyrimidines, see ref. 4.

### Spectra

(a) Ultraviolet.[5]
(b) Infrared.[6]
(c) Proton magnetic resonance.[7]

### Reviews

(a) "The Pyrimidines", "The Chemistry of Heterocyclic Compounds", ed. A. Weissberger, Vol. 16.
(b) "Pyrimidine and its Derivatives", "Heterocyclic Compounds", ed. R. C. Elderfield, Vol. 6, p. 234.
(c) "Pyrimidines", "Chemistry of Carbon Compounds", ed. E. H. Rodd, Vol. IV B, p. 1257.
(d) "Some Aspects of Pyrimidine and Purine Chemistry", B. Lythgoe, *Quart. Rev.*, 1949, **3**, 181.

### References

1. "The Chemistry of Heterocyclic Compounds", ed. A. Weissberger, Vol. 16, p. 31.
2. M. Bachstez, *Ber.*, 1930, **63**, 1000.
3. D. J. Brown, *J. Appl. Chem.*, 1952, **2**, 239.
4. A. R. Katritzky and J. M. Lagowski, *Adv. Heterocyclic Chem.*, 1963, **1**, 371.
5. H. W. Barrett, and R. A. West, *J. Amer. Chem. Soc.*, 1956, **78**, 1612.
6. L. N. Short and H. W. Thompson, *J. Chem. Soc.*, 1952, 168.
7. J. P. Kokko, L. Mandell, and J. H. Goldstein, *J. Amer. Chem. Soc.*, 1962, **84**, 1042.

## HEXAHYDRO-2,4,6-TRIOXOPYRIMIDINE (BARBITURIC ACID)

Barbiturates are prepared by condensing diethyl malonate or its alkylated derivatives, with urea in the presence of a base. Interaction of diethyl malonate with urea in the presence of sodium ethoxide yields hexahydro-2,4,6-trioxopyrimidine (I).[1,2]

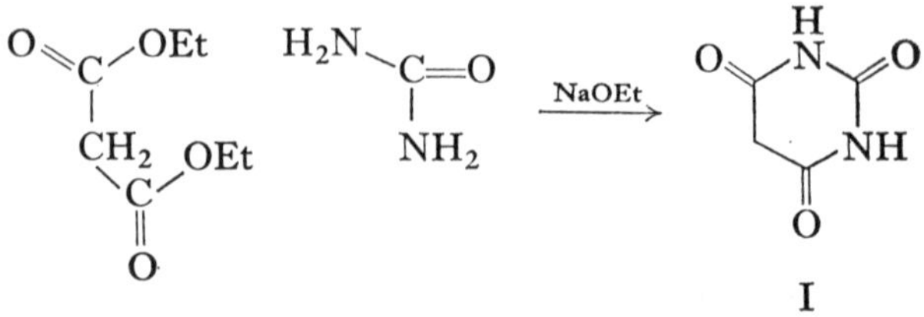


I

### Hexahydro-2,4,6-trioxopyrimidine (I)

In a 250-ml. round-bottomed flask, dissolve freshly cut sodium (1·15 g.) in ethanol (25 ml.). To the solution add, successively, diethyl malonate (8 g.; 7·9 ml.) and a solution of urea (3 g.) in hot ethanol (25 ml.). Heat the resulting mixture on a water-bath at 100° for 1·5 hr., during which time the sodium salt of the product precipitates. Filter the mixture, and dissolve the residual sodium salt in water (100 ml.). Acidify the resulting solution with concentrated hydrochloric acid and filter off the crude product.

Hexahydro-2,4,6-trioxopyrimidine crystallizes from water (ca. 8 ml./g.) as colourless prisms of the dihydrate, which decompose on heating.

Yield = 3·8–4 g. (59·5–62·5%).

For a discussion on the tautomerism of oxopyrimidines, see refs. 3 and 4.

### Spectra

(a) Ultraviolet.[5]
(b) Infrared.[6]

### Reviews

(a) "The History of Barbituric Acid", M. K. Carter, *J. Chem. Educ.*, 1951, **28**, 524.
(b) "Advances in the Chemistry of Barbituric Acid", R. Ya. Levina and F. K. Velichko, *Russ. Chem. Rev.*, 1960, **29**, 437.
(c) "Hydroxy- and Alkoxypyrimidines", "The Chemistry of Heterocyclic Compounds", ed. A. Weissberger, Vol. 16, p. 227.
(d) "Trihydroxypyrimidines", "Chemistry of Carbon Compounds", ed. E. H. Rodd, Vol. IV B, p. 1276.

### References

1. A. Michael, *J. prakt. Chem.*, 1887, **35**, 456.
2. J. B. Dickey and A. R. Gray, *Org. Synth.*, 1943, Coll. Vol. II, 60.
3. J. J. Fox and D. Shugar, *Bull. Soc. chim. belges*, 1952, **61**, 44.
4. A. R. Katritzky and J. M. Lagowski, *Adv. Heterocyclic Chem.*, 1963, **1**, 375.
5. H. G. Mautner and E. M. Clayton, *J. Amer. Chem. Soc.*, 1959, **81**, 6270.
6. Y. Sata, K. Kotera, T. Takabashi, and T. Meshi, *Yakugaku Zasshi*, 1960, **80**, 976.

## 2,5-DIOXOPIPERAZINE

2,5-Dioxopiperazines result from the action of heat on certain α-amino-acids.[1] Heat treatment of glycine (e.g. in boiling glycerol or ethylene glycol) leads to 2,5-dioxopiperazine (I).[2]

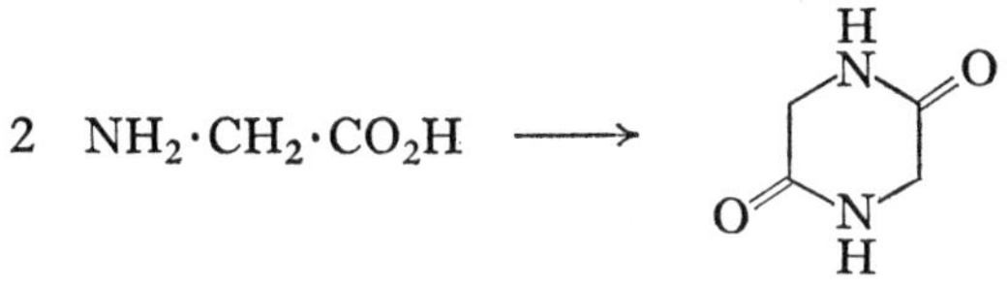

I

### 2,5-Dioxopiperazine (I)

In a 250-ml. round-bottomed flask, heat a solution of glycine (15 g.) in ethylene glycol (90 ml.) under reflux for 0·5 hr. Cool the dark-brown solution in ice–water and filter off the precipitated crude product.

2,5-Dioxopiperazine crystallizes from water (ca. 6 ml./g.) (decolourizing carbon) as white plates, m.p. > 300° (decomp.).

Yield = 4–5 g. (35–44%).

### Spectra

(a) Infrared.[3]
(b) Proton magnetic resonance.[4]

### Reviews

(a) "2,5-Diketopiperazines", "Heterocyclic Compounds", ed. R. C. Elderfield, Vol. 6, p. 435.
(b) "Piperazines (Tetrahydropyrazines)", "Chemistry of Carbon Compounds", ed. E. H. Rodd, Vol. IV B, p. 1336.

### References

1. C. Sannié, *Bull. Soc. chim. France*, 1942, **9**, 487.
2. H. F. Schott, J. B. Larkin, L. B. Rockland, and M. S. Dunn, *J. Org. Chem.*, 1947, **12**, 490.
3. T. Miyazawa, *J. Mol. Spectroscopy*, 1960, **4**, 155.
4. F. A. Bovey and G. V. D. Tiers, *J. Amer. Chem. Soc.*, 1959, **81**, 2870.

## PYRAZINE-2,3-DICARBOXYLIC ACID

Quinoxaline (I) (p. 115) is degraded by potassium permanganate to pyrazine-2,3-dicarboxylic acid (II).[1,2,3]

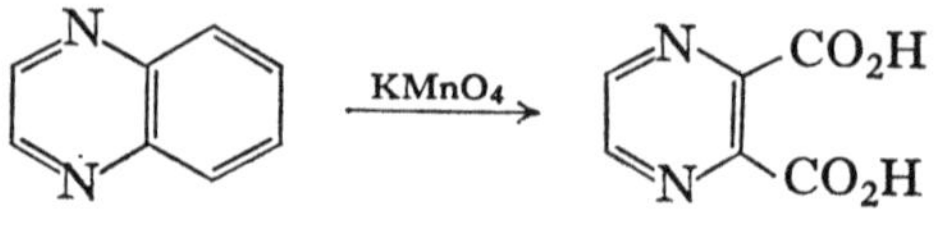


I II

### Pyrazine-2,3-dicarboxylic acid (II)

In a 1-litre 3-necked flask fitted with magnetic stirrer, internal thermometer, 500-ml. dropping-funnel and reflux condenser, suspend quinoxaline (7·25 g.) in water (200 ml.). Heat the suspension to 90°, and with vigorous stirring, add a warm solution of potassium permanganate (52·5 g.) in water (450 ml.). Control the application of heat and rate of addition such that the flask contents boil gently during the addition which takes ca. 45 min. When addition is complete, remove the source of heat, and allow to stir for 15 min. before filtering at the pump. Wash the residual cake of manganese dioxide thoroughly with hot water (100 ml.), and evaporate the combined filtrate and washings under reduced pressure to a volume of ca. 150 ml. Carefully acidify the solution with concentrated hydrochloric acid (27·5 ml.), and continue evaporation to dryness. Grind up the residue, and allow to air-dry overnight.

Separate the product from inorganic material, by heating under reflux with a mixture of acetone (100 ml.) and water ( 8 ml.) for 15 min. Cool to room temperature, filter, and wash the residual inorganic material with acetone (25 ml.). Evaporate the combined filtrate and washings to dryness to obtain the crude product as a brown residue.

Pyrazine-2,3-dicarboxylic acid crystallizes from water (ca. 1·5 ml./g.) (decolourizing carbon), as fawn needles, m.p. 182° (decomp.).

Yield = 5·8–6·3 g. (62–67%).

**Spectra**

Ultraviolet.[4]

**Reviews**

(a) "The Pyrazines", "Heterocyclic Compounds", ed. R. C. Elderfield, Vol. 6, p. 378.

(b) "Pyrazines", "Chemistry of Carbon Compounds", ed. E. H. Rodd, Vol. IV B, 1318.

(c) "The Pyrazines", I. J. Krems and P. E. Spoerri, *Chem. Rev.*, 1947, **40**, 279.

(d) "Kinetics and Mechanism of Oxidation by Permanganate", J. W. Ladbury and (in part) C. F. Cullis, *Chem. Rev.*, 1958, **58**, 403.

**References**

1. J. W. Sausville and P. E. Spoerri, *J. Amer. Chem. Soc.*, 1941, **63**, 3153.
2. I. A. Solomons and P. E. Spoerri, *J. Amer. Chem. Soc.*, 1953, **75**, 679.
3. R. G. Jones and K. C. McLaughlin, *Org. Synth.*, 1963, Coll. Vol. IV, p. 824.
4. S. F. Mason, *J. Chem. Soc.*, 1959, 1247.

## HEXAHYDRO-1,3,5-TRI-*p*-TOLYL-*s*-TRIAZINE

Interaction of *p*-toluidine with formaldehyde can lead to a mixture of di-*p*-toluidinomethane, polymeric methylene *p*-toluidine and trimeric methylene-*p*-toluidine (I), the composition of which depends on the reaction conditions.[1,2] At room temperature, a high yield of hexahydro-1,3,5-tri-*p*-tolyl-*s*-triazine (trimeric methylene *p*-toluidine) (I) is obtained.[3]

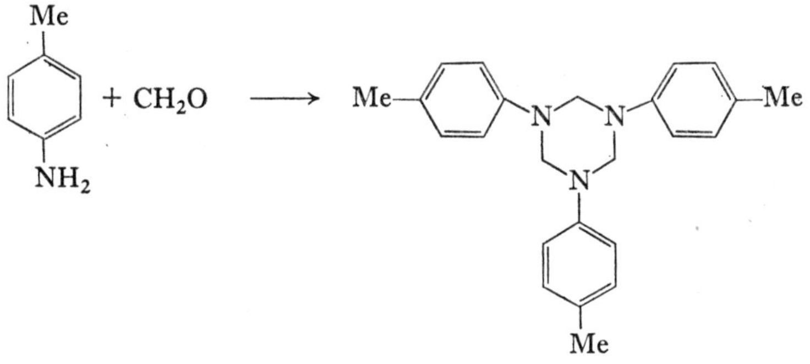


I

### Hexahydro-1,3,5-tri-*p*-tolyl-*s*-triazine (I)

In a 100-ml. conical flask, dissolve *p*-toluidine (10 g.) in ethanol (50 ml.), and at room temperature add 40% aqueous formaldehyde (15 ml.). Allow the mixture to stand at room temperature for 0·5 hr., then filter off the crude product and wash with cold ethanol. Purify the product by *rapidly* extracting with boiling petroleum ether (b.p. 80–100°; 60 ml.). After removal of insoluble high polymer by hot filtration, *cool* the filtrate to room temperature and filter off the product.

Hexahydro-1,3,5-tri-*p*-tolyl-*s*-triazine is obtained as white needles, m.p. 125–127°.

Yield = 7–9 g. (70–90%).

**Spectra**

Ultraviolet.[4]

## Reviews

(a) "Hexahydro-*s*-triazines", "The Chemistry of Heterocyclic Compounds", ed. A. Weissberger, Vol. 13, p. 473.
(b) "*s*-Triazines", "Heterocyclic Compounds", ed. R. C. Elderfield, Vol. 7, p. 627,
(c) "Triazines", "Chemistry of Carbon Compounds", ed. E. H. Rodd, Vol. IV C. p. 1559.

## References

1. C. K. Ingold and H. A. Piggott, *J. Chem. Soc.*, 1923, **123**, 2745.
2. J. G. Miller and E. C. Wagner, *J. Amer. Chem. Soc.*, 1932, **54**, 3698.
3. E. C. Wagner and J. K. Simons, *J. Chem. Educ.*, 1936, **13**, 268.
4. R. Carpignarno, *Ann. Chim. (Rome)*, 1958, **48**, 258.

# 8 BENZODERIVATIVES OF SIX-MEMBERED SYSTEMS CONTAINING MORE THAN ONE HETEROATOM

## 4-CINNOLONE

In the Borsche synthesis,[1] 4-cinnolones result from diazotization of *o*-aminoketones followed by cyclization of the resulting diazonium salt. *o*-Aminoacetophenone (I) and nitrous acid yield 4-cinnolone (II).[2]

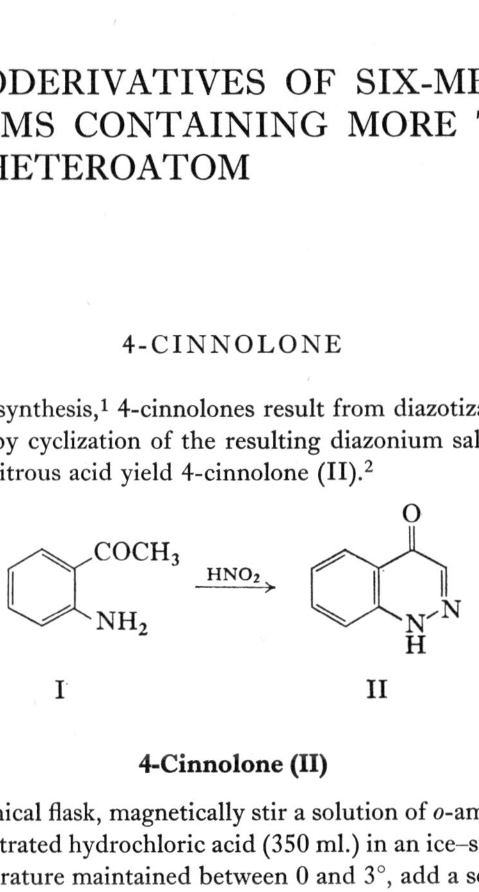


I II

### 4-Cinnolone (II)

In a 500-ml. conical flask, magnetically stir a solution of *o*-aminoacetophenone (4·05 g.) in concentrated hydrochloric acid (350 ml.) in an ice–salt bath, and with the internal temperature maintained between 0 and 3°, add a solution of sodium nitrite (2·28 g.) in water (20 ml.). When the addition is complete, stir at this temperature for 15 min., then heat on a water-bath at 60–75° for 3 hr. Evaporate the solution to dryness under reduced pressure, and dissolve the residual yellow hydrochloride in water (20 ml.). Complete the precipitation of the free base by adding an excess of a saturated solution of sodium acetate, and filter off the crude product.

4-Cinnolone crystallizes from ethanol (ca. 15 ml./g.) (decolourizing carbon) as yellow prisms, m.p. 232–234°.

Yield = 1·8–2·2 g. (41–50%).

For a discussion on the tautomerism of cinnolones, see ref. 3.

### Spectra

(a) Ultraviolet.[4]
(b) Infrared.[5]

### Reviews

(a) "4-Hydroxycinnolines", "The Chemistry of Heterocyclic Compounds", ed. A. Weissberger, Vol. 5, p. 16.
(b) "Cinnolines and Related Compounds", "Heterocyclic Compounds", ed. R. C. Elderfield, Vol. 6, p. 136.
(c) "Cinnolines", "Chemistry of Carbon Compounds", ed. E. H. Rodd, Vol. IV B, p. 1217.
(d) "The Chemistry of Cinnolines", N. J. Leonard, *Chem. Rev.*, 1945, **37**, 269.

### References

1. W. Borsche and A. Herbert, *Annalen*, 1941, **546**, 293.
2. J. R. Keneford and J. C. E. Simpson, *J. Chem. Soc.*, 1947, 917.
3. A. R. Katritzky and J. M. Lagowski, *Adv. Heterocyclic Chem.*, 1963, **1**, 364.
4. J. M. Hearn, R. A. Morton, and J. C. E. Simpson, *J. Chem. Soc.*, 1951, 3318.
5. S. F. Mason, *J. Chem. Soc.*, 1957, 4874.

## QUINAZOLINE

Formamide condenses with *o*-nitrobenzaldehyde in the presence of hydrogen chloride to give α,α-diformamido-*o*-nitrotoluene (I).[1] Reduction of the nitrodiamide with iron and hydrochloric acid yields quinazoline (II).[1]

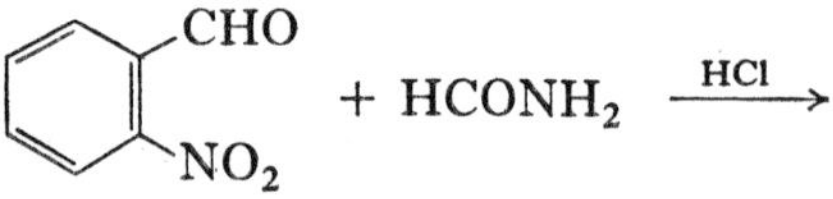


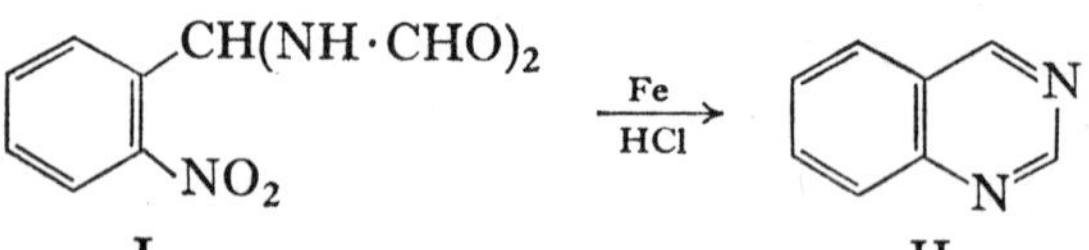


I II

### α,α-Diformamido-*o*-nitrotoluene (I)

In a 100-ml. round-bottomed flask fitted with an internal thermometer, heat a solution of *o*-nitrobenzaldehyde (10 g.) in formamide (20 g.; 17·6 ml.) on a water-bath. When the temperature reaches 80°, saturate the solution with hydrogen chloride gas for ca. 20 min.

**Note**

Hydrogen chloride gas is conveniently generated by dropping concentrated sulphuric acid onto ammonium chloride contained in a Büchner flask.

Allow the resulting pale-yellow semi-solid mixture to stand overnight at room temperature, then stir the solid mass with ethanol (30 ml.) until a finely divided yellow precipitate remains. Filter the mixture, and wash the residue with 50% aqueous ethanol (2 × 60 ml.).

α,α-Diformamido-*o*-nitrotoluene is obtained as a pale-yellow crystalline solid, m.p. 170–172° (decomp.), sufficiently pure for direct use in the next stage.

Yield = 9–9·6 g. (61–65%).

## Quinazoline (II)

In a 250-ml. conical flask, mechanically stir a suspension of reduced iron powder (5 g.) in water (100 ml.). Add concentrated hydrochloric acid (1 ml.), followed by α,α-diformamido-*o*-nitrotoluene (4 g.). When addition is complete (ca. 5 min.), heat the mixture with constant stirring on a water-bath at 100° for 3 hr. Cool, filter off the excess of iron powder, and to the filtrate add excess of solid sodium carbonate. Transfer the mixture into a separating funnel, and extract with ether (2 × 30 ml.). Combine the extracts, dry ($MgSO_4$), and evaporate off the ether to obtain the product as a yellow oil that solidifies on cooling.

Quinazoline crystallizes (with difficulty) from petroleum ether (b.p. 40–60°; ca. 20 ml./g.) as pale-yellow prisms, m.p. 41–42°.

Yield = 1·5–1·8 g. (64·5–77%).

**Spectra**

Ultraviolet.[2]

**Reviews**

(a) "Quinazolines", W. L. F. Armarego, *Adv. Heterocyclic Chem.*, 1963, **1**, 253.
(b) "The Chemistry of Quinazoline", "Heterocyclic Compounds", ed. R. C. Elderfield, Vol. 6, p. 324.
(c) "Quinazolines", "Chemistry of Carbon Compounds", ed. E. H. Rodd, Vol. IV B, p. 1299.

**References**

1. K. Adachi, *J. Pharm. Soc., Japan*, 1955, **75**, 1423. (*Chem Abs.*, 1956, **50**, 10105).
2. A. R. Osborn and K. Schofield, *J. Chem. Soc.*, 1956, 4191.

## 3-AMINO-3,4-DIHYDRO-2-METHYL-4-OXOQUINAZOLINE

Treatment of methyl *N*-acetylanthranilate (I) with hydrazine hydrate yields 3-amino-3,4-dihydro-2-methyl-4-oxoquinazoline (II).[1]

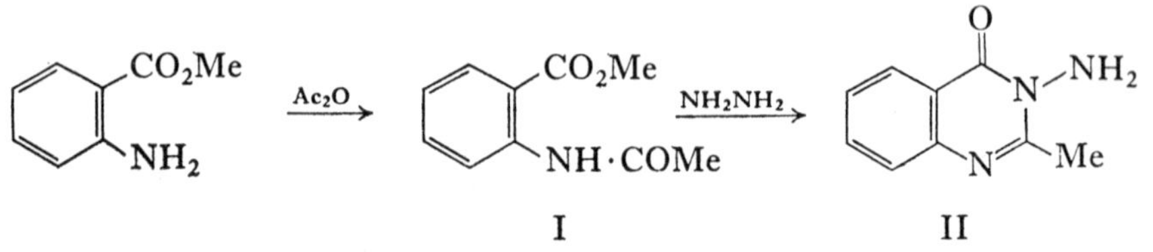

### Methyl *N*-acetylanthranilate (I)

In a 100-ml. round-bottomed flask, heat a solution of methyl anthranilate (3·2 g.; 2·7 ml.) in acetic anhydride (12 ml.) under reflux for 15 min. Cool the solution, pour into cold water (50 ml.), containing a drop of pyridine, and stir until the oil solidifies. Filter off the product, wash with cold water (4 × 50 ml.), and dry at 100°.

Methyl *N*-acetylanthranilate crystallizes from ethanol (ca. 6 ml./g.) as colourless needles, m.p. 100–101°.

Yield = 2·8–3·2 g. (68·5–78%).

### 3-Amino-3,4-dihydro-2-methyl-4-oxoquinazoline (II)

In a 100-ml. round-bottomed flask, heat a solution of hydrazine hydrate (10 ml.) and methyl *N*-acetylanthranilate (2 g.) in ethanol under reflux for 2 hr. Cool the mixture, and stir into cold water (50 ml.). Filter off the crude product, wash with cold water and dry at 100°.

3-Amino-3,4-dihydro-2-methyl-4-oxoquinazoline crystallizes from ethanol (ca. 6 ml./g.) as colourless needles, m.p. 152–153°.

Yield = 1·5–1·6 g. (82·5–88%).

**Reviews**

(a) "Synthesis in the 3-Aminoquinazol-4-one and 3-Aminobenzotriazin-4-one series", S. Petersen, H. Herlinger, E. Tietze, and W. Siefken, *Angew. Chem.* (*Internat. edn.*), 1963, **2**, 24.
(b) "The Chemistry of Quinazoline", "Heterocyclic Compounds", ed. R. C. Elderfield, Vol. 6, p. 324.
(c) "Quinazolines", W. L. F. Armarego, *Adv. Heterocyclic Chem.*, 1963, **1**, 253.
(d) "Hydrogenated Quinazolines", "Chemistry of Carbon Compounds", ed. E. H. Rodd, Vol. IV B, p. 1315.

### Reference

1. G. Heller, *J. prakt. Chem.*, 1925, **111**, 40.

## QUINOXALINE

α-Dicarbonyl compounds condense with *o*-diamines to yield quinoxalines.[1] Glyoxal and *o*-phenylenediamine yield quinoxaline (I), but the yield is improved by use of glyoxal–sodium hydrogen sulphite.[2]

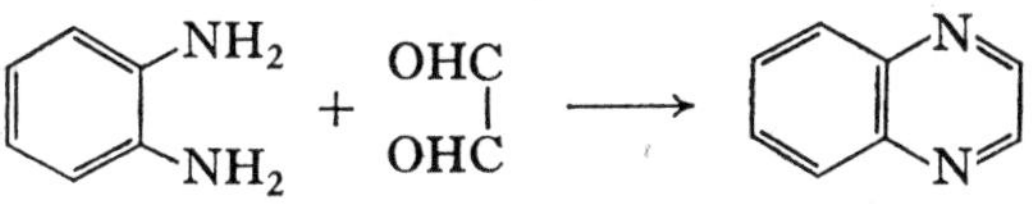


I

### Quinoxaline (I)

Prepare a solution of glyoxal–sodium hydrogen sulphite by adding 30% aqueous glyoxal (18 ml.) to a solution of sodium metabisulphite (20·9 g.) in water (100 ml.). After heating to ca. 70°, add this solution to a solution of *o*-phenylenediamine (10·8 g.) in hot water (150 ml.), and allow the resulting mixture to stand for 15 min. Cool to room temperature and basify by adding a solution of sodium carbonate (45 g.) in water (100 ml.), then extract the liberated product with ether (3 × 50 ml.). Combine the extracts, dry ($MgSO_4$), and after evaporating off the ether, distil the residue (which may solidify) under water-pump vacuum.

Quinoxaline is obtained as an almost colourless liquid, b.p. 109–111°/12 mm., which solidifies on cooling to an ice-like solid, m.p. 29–30°.

Yield = 9·6–10·4 g. (74–80%).

### Spectra

(a) Ultraviolet.[3]
(b) Proton magnetic resonance.[4]

### Reviews

(a) "Quinoxalines", "The Chemistry of Heterocyclic Compounds", ed. A. Weissberger, Vol. 5, p. 203.
(b) "The Quinoxalines", "Heterocyclic Compounds", ed. R. C. Elderfield, Vol. 6, p. 455.
(c) "Quinoxalines", "Chemistry of Carbon Compounds", ed. E. H. Rodd, Vol. IV B, p. 1345.
(d) "Recent Advances in Quinoxaline Chemistry", G. W. H. Cheeseman, *Adv. Heterocyclic Chem.*, 1963, **2**, 203.

**References**

1. O. Hinsberg, *Annalen*, 1887, **237**, 327.
2. J. H. Billman and J. L. Rendall, *J. Amer. Chem. Soc.*, 1944, **66**, 540.
3. G. W. H. Cheeseman, *J. Chem. Soc.*, 1958, 108.
4. S. Matsuura and T. Goto, *J. Chem. Soc.*, 1963, 1773.

## 3,4-DIHYDRO-1-HYDROXY-4-OXOPHTHALAZINE

Treatment of phthalic anhydride with hydrazine hydrate gives the cyclic dihydrazide, 3,4-dihydro-1-hydroxy-4-oxophthalazine (I).[1]

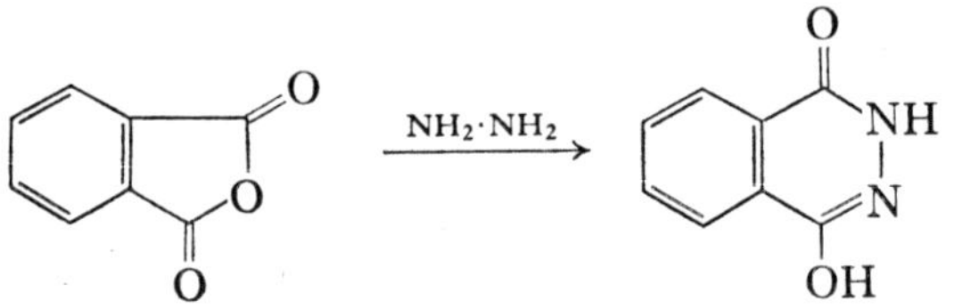


I

### 3,4-Dihydro-1-hydroxy-4-oxophthalazine (I)

In a 100-ml. round-bottomed flask, heat under reflux a solution of hydrazine hydrate (1 ml.) and phthalic anhydride (3 g.) in ethanol (25 ml.) on a water-bath at 100° for 0·5 hr. Cool the reaction mixture, filter off the product and wash with petroleum ether (b.p. 40–60°; 3 × 40 ml.).

3,4-Dihydro-1-hydroxy-4-oxophthalazine is obtained as a white crystalline solid, m.p. 334–336°, and is virtually pure.

Yield = 2·6–2·9 g. (79–88·5%).

For a discussion on the tautomerism of phthalazones see ref. 2.

**Spectra**

(a) Ultraviolet.[3]
(b) Infrared.[3]

**Reviews**

"Hydroxyphthalazines", see review references on p. 117.

**References**

1. T. Curtius and H. A. Foersterling, *J. prakt. Chem.*, 1895, **51**, 376.
2. A. R. Katritzky and J. M. Lagowski, *Adv. Heterocyclic Chem.*, 1963, **1**, 366.
3. J. A. Elvidge and A. P. Redman, *J. Chem. Soc.*, 1960, 1710.

## 3,4-DIHYDRO-1-METHYL-4-OXOPHTHALAZINE

*o*-Dicarbonyl compounds react with hydrazine to give phthalazines. Interaction of *o*-acetylbenzoic acid (I) with hydrazine hydrate in refluxing ethanol gives 3,4-dihydro-1-methyl-4-oxophthalazine (II) in excellent yield.[1]

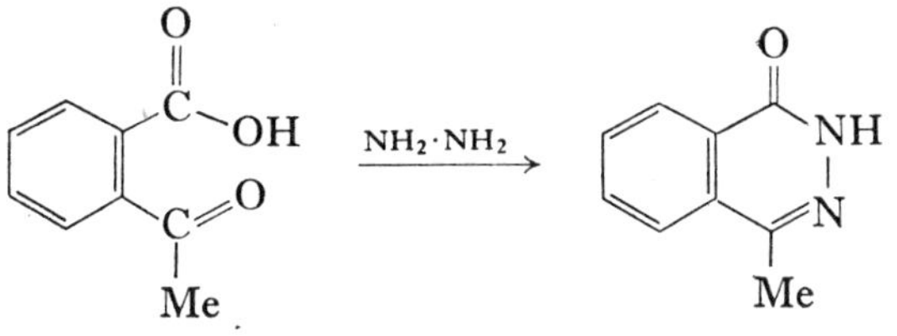


I II

### 3,4-Dihydro-1-methyl-4-oxophthalazine (II)

In a 100-ml. round-bottomed flask, heat a solution of hydrazine hydrate (4 ml.) and *o*-acetylbenzoic acid (4·1 g.) in ethanol (25 ml.) under reflux for 0·5 hr. Cool the mixture, filter and wash the residue with ice-cold ethanol ($2 \times 25$ ml.).

3,4-Dihydro-1-methyl-4-oxophthalazine is obtained as a white crystalline product, m.p. 220–222°, and is virtually pure.

Yield = 2·8–3·2 g. (80–90%).

For a discussion on the tautomerism of phthalazones, see refs. 2 and 3.

### Spectra

Infrared.[4]

### Reviews

(a) "The Chemistry of the Phthalazines", W. R. Vaughan, *Chem. Rev.*, 1948, **43**, 447.
(b) "Phthalazines", "The Chemistry of Heterocyclic Compounds", ed. A. Weissberger, Vol. 5, p. 1.
(c) "Phthalazine and its derivatives", "Heterocyclic Compounds", ed. R. C. Elderfield, Vol. 6, p. 186.
(d) "Hydroxyphthalazines and Dihydrophthalazines", "Chemistry of Carbon Compounds", ed. E. H. Rodd, Vol. IV B, p. 1244.

### References

1. S. Gabriel and A. Neumann, *Ber.*, 1893, **26**, 521, 705.
2. F. M. Rowe and A. T. Peters, *J. Chem. Soc.*, 1933, 1331.
3. A. R. Katritzky and J. M. Lagowski, *Adv. Heterocyclic Chem.*, 1963, **1**, 366.
4. S. F. Mason, *J. Chem. Soc.*, 1957, 4874.

## DIBENZO[*a*,*c*]PHENAZINE

*o*-Quinones readily condense with *o*-phenylenediamines to give phenazine derivatives (cf. preparation on p. 115). Interaction of *o*-phenylenediamine and phenanthraquinone (I) in glacial acetic acid gives dibenzo[*a*,*c*]phenazine (II).[1]

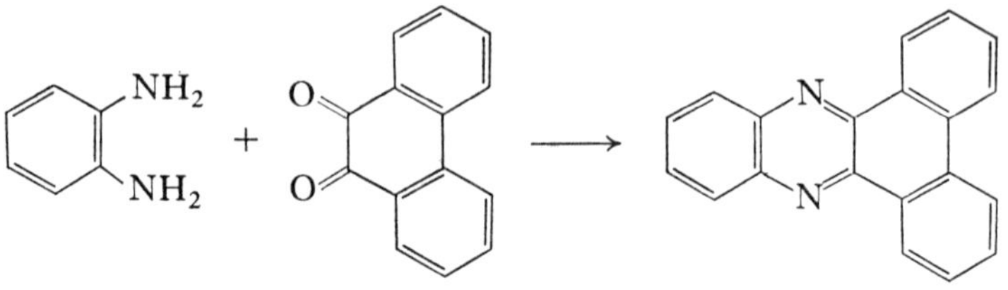

I II

### Dibenzo[*a*,*c*]phenazine (II)

To a solution of *o*-phenylenediamine (1·1 g.) in ethanol (10 ml.), contained in a 100-ml. conical flask, add a solution of phenanthraquinone (2·1 g.) in warm glacial acetic acid (70 ml.). Heat the resulting mixture on a water-bath at 100° for 0·5 hr., then cool and filter off the yellow product. Wash the residue with cold ethanol (2 × 20 ml.), and dry at 100°.

Dibenzo[*a*,*c*]phenazine crystallizes from glacial acetic acid (ca. 10 ml./g.) as pale yellow needles, m.p. 221–223°.

Yield = 2·2–2·5 g. (77–89·5%).

### Spectra

Ultraviolet.[2]

### Reviews

(a) "Benzo[*a*,*c*]phenazines", "The Chemistry of Heterocyclic Compounds", ed. A. Weissberger, Vol. II, p. 363.
(b) "Phenazines", "Heterocyclic Compounds", ed. R. C. Elderfield, Vol. 6, p. 624.
(c) "Benzo- and Dibenzophenazines", "Chemistry of Carbon Compounds", ed. E. H. Rodd, Vol. IV B, p. 1376.

### References

1. O. Hinsberg, *Annalen*, 1887, **237**, 327.
2. R. Epsztein, *Mém. Services chim. État*, 1951, **36**, 353.

## 3,4-DIHYDRO-4-OXO-1,2,3-BENZOTRIAZINE

METHOD 1

Diazotization of anthranilohydrazide (I) in hydrochloric acid solution gives a mixture of 3,4-dihydro-4-oxo-1,2,3-benzotriazine (II) and *o*-aminobenzazide (III).[1] 3,4-Dihydro-4-oxo-1,2,3-benzotriazine arises from internal coupling of the diazonium cation with the adjacent hydrazide group, followed by deamination.

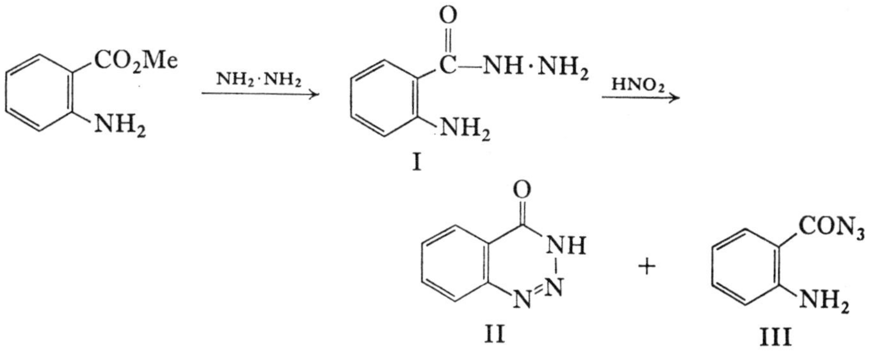


## Anthranilohydrazide (I)

In a 100-ml. round-bottomed flask, heat under reflux a solution of hydrazine hydrate (12 ml.) and methyl anthranilate (10 g.; 8·5 ml.) in ethanol (10 ml.) for 2 hr. Cool the mixture, filter off the crystalline product and wash with ice-cold ethanol (2 × 20 ml.).

Anthranilohydrazide is obtained as a brown solid, m.p. 121–122°, sufficiently pure for direct use in the next stage.

Yield = 8·5–9 g. (85–90%).

## 3,4-Dihydro-4-oxo-1,2,3-benzotriazine (II)

To an ice-cold, stirred solution of anthranilohydrazide (7·5 g.) in concentrated hydrochloric acid (15 ml.) and water (15 ml.), contained in a 250-ml. conical flask, add a solution of sodium nitrite (3·9 g.) in water (10 ml.). Stir the mixture for 5 min., then filter off the white solid and retain the filtrate which contains *o*-aminobenzazide. Wash the residue with cold water (2 × 50 ml.) and dry at 100°.

3,4-Dihydro-4-oxo-1,2,3-benzotriazine crystallizes from ethanol (ca. 12 ml./g.) as pale-yellow prisms, m.p. 210° (decomp.)

Yield = 1·6–1·9 g. (20–23·5%).

## Anthranilic acid azide (III)

Basify the filtrate from the above experiment by adding an excess of solid sodium carbonate. Stir the mixture vigorously and filter off the precipitate. Wash the crude product with water (3 × 40 ml.), and dry in a vacuum desiccator over concentrated sulphuric acid.

### Note

Nearly all azides decompose violently when heated alone either in the solid or the liquid state. The prepared azide must *not*, therefore, be dried over a flame or in an oven.

Anthranilic acid azide is obtained as a brown solid, m.p. 78–80° (decomp.), sufficiently pure for direct use in the preparation given on p. 43.

Yield = 1·8–2·5 g. (22·5–31%).

### Note

Anthranilic acid azide undergoes slow hydrolysis to the isocyanate, which is a strong lachrymator. The dry azide should, therefore, be kept in a well stoppered bottle, preferably in a cool place.

### Reviews

(a) "Synthesis in the 3-Aminoquinazol-4-one and 3-Aminobenzotriazin-4-one series", S. Petersen, H. Herlinger, E. Tietze, and W. Siefken, *Angew. Chem.* (*Internat. edn.*), 1963, **2**, 24.

(b) "Benzotriazinones", see review references on p. 121.

### Reference

1. G. Heller and A. Siller, *J. prakt. Chem.*, 1927, **116**, 9.

## 3,4-DIHYDRO-4-OXO-1,2,3-BENZOTRIAZINE

### METHOD 2

3,4-Dihydro-4-oxo-1,2,3-benzotriazine (III) results from the neutralization of diazotized methyl anthranilate with ammonium hydroxide.[1] The initially formed *o*-diazo-ester (I) is converted into the diazo-amide (II),† which undergoes rapid internal coupling to yield the triazine (III).

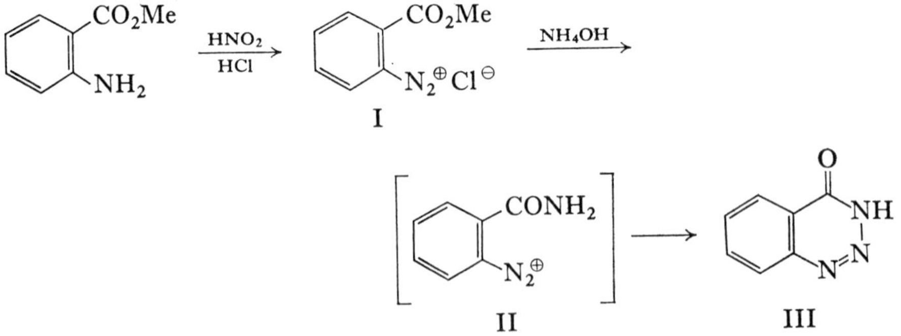


† Diazotization of *o*-aminobenzamide gives 3,4-dihydro-4-oxo-1,2,3-benzotriazine in quantitative yield.

### 3,4-Dihydro-4-oxo-1,2,3-benzotriazine (III)

To an ice-cold solution of methyl anthranilate (7·6 g.; 8 ml.) in concentrated hydrochloric acid (20 ml.) and water (25 ml.), contained in a 250-ml. beaker, add slowly a solution of sodium nitrite (4 g.) in water (20 ml.). Stir the resulting diazonium solution for 5 min. at 0°, then add dropwise concentrated ammonium hydroxide until the solution is just alkaline. Stir the mixture for 5 min., and then filter off the brown precipitate. Wash the residue with ice-cold ethanol (4 × 25 ml.), and dry at 100°.

3,4-Dihydro-4-oxo-1,2,3-benzotriazine crystallizes from ethanol (ca. 12 ml./g.) as pale-yellow prisms, m.p. 210° (decomp.).

Yield = 3·6–4 g. (48·5–54%).

#### Spectra

Ultraviolet.[2]

#### Reviews

(a) "Synthesis in the 3-Aminoquinazol-4-one and 3-Aminobenzotriazin-4-one series", S. Petersen, H. Herlinger, E. Tietze and W. Siefken, *Angew. Chem.*, (*Internat. edn.*), 1963, **2**, 24.
(b) "1,2,3-Benzotriazines. Hydroxy- and oxo-derivatives", "The Chemistry of Heterocyclic Compounds", ed. A. Weissberger, Vol. 10, p. 13.
(c) "The 1,2,4- and 1,2,3-Triazines", "Heterocyclic Compounds", ed. R. C. Elderfield, Vol. 7, p. 720.
(d) "1,2,3-Benzotriazines (Phenotriazines)", "Chemistry of Carbon Compounds", ed. E. H. Rodd, Vol. IV C, p. 1559.

#### References

1. E. Zacharias, *J. prakt. Chem.*, 1891, **43**, 446.
2. D. Buckley and M. S. Gibson, *J. Chem. Soc.*, 1956, 3242.

## 2-ETHYL-4-OXO-3,1-BENZOXAZINE

*N*-Acylated anthranilic acids are cyclized by boiling acetic anhydride to substituted 4-oxo-3,1-benzoxazines. Under these conditions *N*-propionylanthranilic acid (I) yields 2-ethyl-4-oxo-3,1-benzoxazine (II).[1]

### *N*-Propionylanthranilic acid (I)

In a 100-ml. round-bottomed flask, heat a solution of anthranilic acid (6·85 g.) in propionic anhydride (20 ml.) under reflux for 0·5 hr. Cool and pour the resulting solution into cold water (100 ml.) containing a drop of pyridine. Stir the

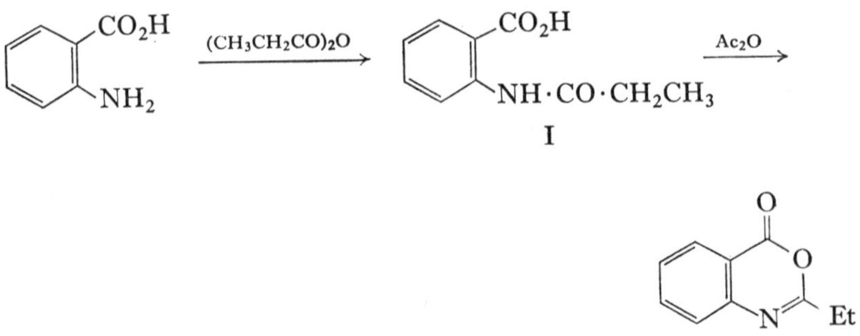


II

mixture until the product solidifies, then filter and wash the residue with cold water (3 × 50 ml.) and dry at 100°.

*N*-Propionylanthranilic acid is obtained as a white solid, m.p. 112–114°, sufficiently pure for direct use in the next stage.

Yield = 5–5·5 g. (52–57%).

## 2-Ethyl-4-oxo-3,1-benzoxazine (II)

In a 100-ml. round-bottomed flask, heat a solution of *N*-propionylanthranilic acid (5 g.) in acetic anhydride (40 ml.) under reflux for 1 hr. Distil off the excess of acetic anhydride (ca. 37 ml.), and stir the residual yellow oil with petroleum ether (b.p. 40–60°; ca. 30 ml.) until solidification occurs. Filter off the crude product and wash successively with ice-cold ethanol (2 × 25 ml.) and petroleum ether (b.p. 40–60°; 3 × 30 ml.).

2-Ethyl-4-oxo-3,1-benzoxazine (II) crystallizes from ethyl acetate (ca. 4 ml./g.) as white needles, m.p. 85–86°.

Yield = 2·8–3·3 g. (61·5–73%).

### Reviews


(a) "3,1-Benzoxazines", "The Chemistry of Heterocyclic Compounds", ed. A. Weissberger, Vol. 17, p. 359.

(b) "The Benzoxazines", "Heterocyclic Compounds", ed. R. C. Elderfield, Vol. 6, p. 564.

(c) "3,1,4-Benzoxazines", "Chemistry of Carbon Compounds", ed. E. H. Rodd, Vol. IV C, p. 1486.


### Reference


1. E. C. Wagner and D. T. Zentmyer, *J. Org. Chem.*, 1949, **14**, 967.

# 2,3-DIHYDRO-3-OXO-1,4-BENZOXAZINE (BENZO-3-MORPHOLONE)

## METHOD 1

Reduction of ethyl *o*-nitrophenoxyacetate (I), obtained from the interaction of potassium *o*-nitrophenate and ethyl chloroacetate,[1] yields the corresponding amino-compound (II), which undergoes rapid cyclization to give 2,3-dihydro-3-oxo-1,4-benzoxazine (III),[2,3]

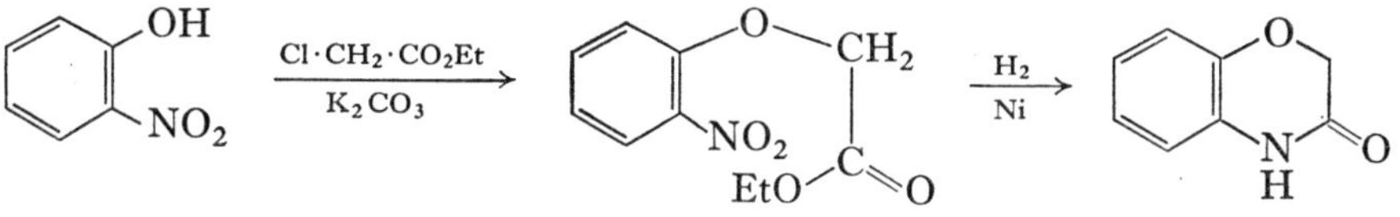


I II

### Ethyl *o*-nitrophenoxyacetate (I)

In a 250-ml. 2-necked, round-bottomed flask, equipped with reflux condenser and stirrer, dissolve *o*-nitrophenol (5·6 g.) in acetone (50 ml.). To the solution add anhydrous potassium carbonate (6 g.), and a solution of ethyl chloroacetate (4·8 g.; 4·2 ml.) in acetone (10 ml.), and heat the constantly stirred mixture under reflux overnight. Cool the reaction mixture, filter off the inorganic material, and evaporate off the acetone from the filtrate. Dissolve the residue in ether (100 ml.), and transfer the solution into a separating funnel. Wash the ethereal solution with 2N sodium hydroxide (2 × 30 ml.), in order to remove unreacted nitrophenol, and then with cold water (2 × 50 ml.). Dry the ethereal solution ($MgSO_4$), and evaporate off the ether to obtain the crude product as a pale-yellow oil that solidifies on cooling.

Ethyl *o*-nitrophenoxyacetate is obtained as a pale-yellow crystalline solid, m.p. 44–45°, sufficiently pure for direct use in the next stage.

Yield = 4·4–4·8 g. (48·5–53%).

### Preparation of Raney nickel catalyst

(This preparation must be carried out in a fume-cupboard in the absence of naked flames, since substantial quantities of hydrogen gas are evolved).

In a 5-litre beaker, suspend 50–50 Raney nickel alloy (25 g.) in a mixture of water (125 ml.) and 5N sodium hydroxide (5 ml.). Mechanically stir the mixture, and when the initial reaction has ceased, add 10N sodium hydroxide (75 ml.). Immediately pull down the fume-cupboard hood and continue stirring until the vigorous exothermic reaction is complete and the mixture has cooled (ca. 0·5 hr.).

Wash the black suspension of finely divided nickel by decantation with cold water (5 × 600 ml.), and then with methanol (100 ml.). Transfer the nickel to a 250-ml. beaker and again wash the suspension by decantation with methanol (3 × 50 ml.). Store the final product under methanol (60 ml.) until required.

**Note**

Freshly prepared Raney nickel catalyst is pyrophoric and must not be allowed to dry either on a filter paper or in the presence of anything combustible. It is also for this reason that waste or residual nickel catalyst should be destroyed by dissolving in 6N hydrochloric acid, and not thrown down the drain or left on the open bench.

### 2,3-Dihydro-3-oxo-1,4-benzoxazine (II)

To a solution of ethyl *o*-nitrophenoxyacetate (2 g.) in benzene (80 ml.) contained in a 250-ml. hydrogenation flask, add a settled suspension (3 ml.) of the prepared Raney nickel catalyst in methanol. Fit the flask to a standard atmospheric hydrogenation apparatus[4] and shake in a hydrogen atmosphere until the theoretical quantity (640 ml.) of hydrogen gas has been absorbed. Allow the suspension of nickel to settle, decant off the benzene–methanol solution and destroy the residual nickel catalyst with 6N hydrochloric acid as instructed above. Evaporate off the solvent to obtain the product as a crystalline residue.

2,3-Dihydro-3-oxo-1,4-benzoxazine crystallizes from ethanol (*ca.* 8 ml./g.) (decolourizing carbon), as colourless prisms, m.p. 164–165°.

Yield = 1–1·2 g. (75·5–90·5%).

METHOD 2

*o*-Aminophenol is readily *N*-acylated[5] in benzene solution by chloroacetyl chloride. Cyclization of the resulting *N*-chloroacetyl compound (I) to 2,3-dihydro-3-oxo-1,4-benzoxazine (II) is effected by base.

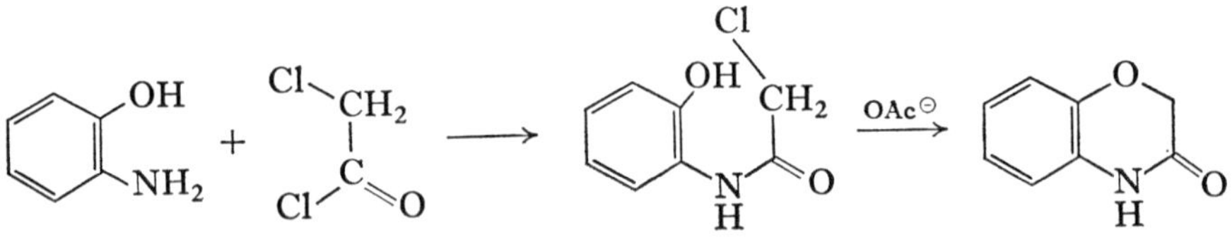


I II

### *o*-(*N*-Chloroacetamido)phenol (I)

To a suspension of *o*-aminophenol (5·5 g.) in sodium-dried benzene (100 ml.), contained in a 250-ml. round-bottomed flask, cautiously add chloroacetyl chloride (5·7 g.). Heat the mixture under reflux on a water-bath at 100° until the suspension of *o*-aminophenol disappears and a clear brown solution is obtained (ca. 2 hr.). Rapidly filter the hot solution, and allow to cool before filtering off the product.

*o*-(*N*-Chloroacetamido)phenol is obtained as a mauve solid, m.p. 137–138°, sufficiently pure for direct use in the next stage.

Yield = 6·5–7·2 g. (69·5–77%).

### 2,3-Dihydro-3-oxo-1,4-benzoxazine (II)

To a solution of *o*-(*N*-chloroacetamido)phenol (3·7 g.) in ethanol, contained in a 250-ml. round-bottomed flask, add anhydrous potassium acetate (5 g.), and heat the mixture under reflux on a water-bath at 100° for 1 hr. Filter the hot mixture, then cool and precipitate the product from the filtrate by addition of cold water. Filter off the crude product, wash with cold water (2 × 30 ml.) and dry at 100°.

2-3-Dihydro-3-oxo-1,4-benzoxazine (II) crystallizes from ethanol (ca. 8 ml./g.) as colourless prisms, m.p. 165°.

Yield = 1·8–2·2 g. (60·5–74%).

### Spectra

Infrared.[6]

### Reviews

(a) "1,4-Benzoxazines", "The Chemistry of Heterocyclic Compounds", ed. A. Weissberger, Vol. 17, p. 394.
(b) "The Benzoxazines", "Heterocyclic Compounds", ed. R. C. Elderfield, Vol. 6, p. 588.
(c) "Benzo-1,4-oxazines and their Hydro-derivatives", "Chemistry of Carbon Compounds", ed. E. H. Rodd, Vol. IV C, p. 1488.

### References

1. T. H. Minton and H. Stephen, *J. Chem. Soc.*, 1922, 121, 1591.
2. G. Newbery and M. A. Phillips, *J. Chem. Soc.*, 1928, 3046.
3. W. G. Christiansen, *J. Amer. Chem. Soc.*, 1926, **48**, 460.
4. A. I. Vogel, "Practical Organic Chemistry", Longmans, Green & Co., London, 3rd edn., 1956, p. 849.
5. O. Aschan, *Ber.*, 1887, **20**, 1524.
6. D. G. O'Sullivan and P. W. Sadler, *J. Chem. Soc.*, 1957, 2916.

## 2,4-DINITROPHENOXAZINE

Picryl chloride (I) condenses in neutral solution with *o*-aminophenol to yield 2′-hydroxy-2,4,6-trinitrodiphenylamine (II). This on treatment with base eliminates nitrous acid to give 2,4-dinitrophenoxazine (III).[1]

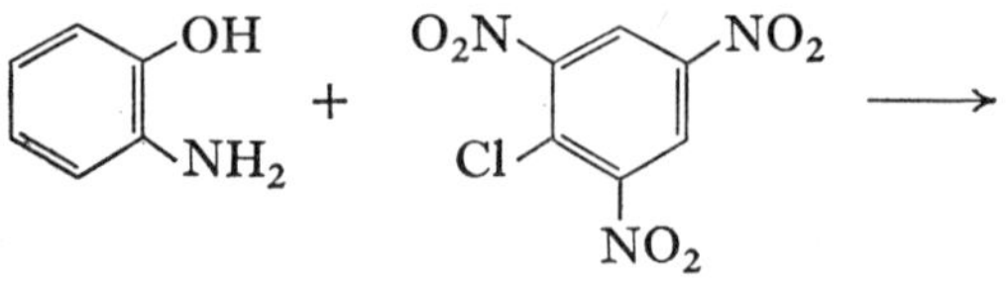


I

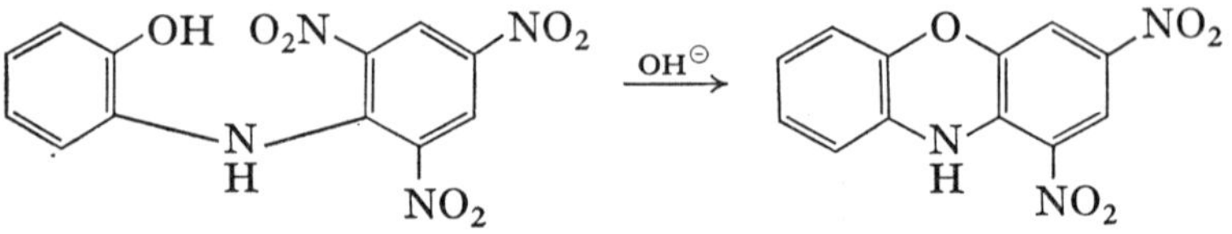


II III

### 2′-Hydroxy-2,4,6-trinitrodiphenylamine (II)

To a solution of *o*-aminophenol (3·3 g.) in ethanol (15 ml.), contained in a 250-ml. round-bottomed flask, add anhydrous sodium acetate (5 g.), and a suspension of picryl chloride (7·4 g.) in ethanol (50 ml.). Heat the mixture under reflux for 2 hr. with occasional shaking, then cool and filter off the precipitated product.

2′-Hydroxy-2,4,6-trinitrodiphenylamine is obtained as deep-brown needles, m.p. 175°, sufficiently pure for direct use in the next stage.

Yield = 8–9 g. (82·5–93%).

### 2,4-Dinitrophenoxazine (III)

In a 100-ml. conical flask, heat a solution of 2′-hydroxy-2,4,6-trinitrodiphenylamine (2 g.) in 0·05N sodium hydroxide (50 ml.) on a water-bath at 100° for 0·5 hr. Cool the reaction mixture, filter off the crystalline precipitate and dry at 100°.

2,4-Dinitrophenoxazine crystallizes from glacial acetic acid (ca. 15 ml./g.) as lustrous red-brown needles, m.p. 215–217°.

Yield = 1·5–1·6 g. (88–94%).

**Note**

The authors found that cyclization to the phenoxazine may occur during the preparation of the substituted diphenylamine (II). In this case the product is, of course, insoluble in sodium hydroxide, and the second stage of the synthesis is unnecessary.

**Reviews**

(a) "Phenoxazines", "The Chemistry of Heterocyclic Compounds", ed. A. Weissberger, Vol. 17, p. 408.
(b) "Phenoxazines", "Heterocyclic Compounds", ed. R. C. Elderfield, Vol. 6, p. 624.
(c) "Phenoxazines", "Chemistry of Carbon Compounds", ed. E. H. Rodd, Vol. IV C, p. 1491.

**Reference**

1. G. S. Turpin, *J. Chem. Soc.*, 1891, **59**, 723.

## PHENOTHIAZINE

Phenothiazine (II) is prepared by heating diphenylamine (I) with sulphur at 200° in the presence of an iodine catalyst.[1]

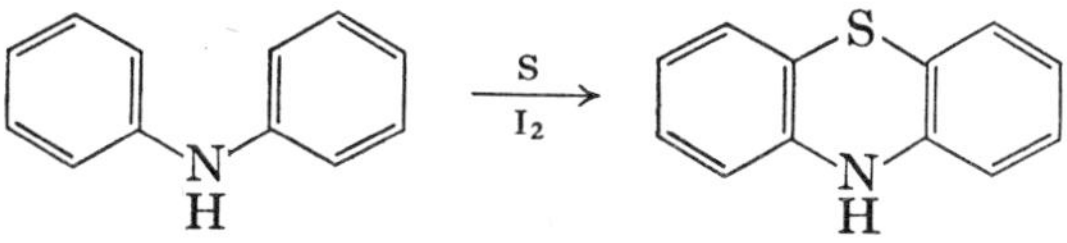

I II

**Phenothiazine (II)**

**Note**

It is advisable to carry out this preparation in an efficient fume cupboard since hydrogen sulphide is evolved during the reaction.

In a 100-ml. round-bottomed flask, heat an intimate mixture of sulphur (3·2 g.), diphenylamine (8·4 g.), and iodine (0·15 g.) in a Wood's metal-bath at 190–200° for 0·5 hr. Cool the molten mixture, and grind the residual mass to a fine powder with a pestel and mortar.

Phenothiazine crystallizes from ethanol (ca. 6 ml./g.) (decolourizing carbon) as yellow plates, m.p. 184–185°.

Yield = 8·5–9 g. (86–91%).

**Spectra**

Ultraviolet.[2]

**Reviews**

(a) "The Chemistry of Phenothiazine", S. P. Massie, *Chem. Rev.*, 1954, **54**, 797.
(b) "Phenothiazines", "Heterocyclic Compounds", ed. R. C. Elderfield, Vol. 6, p. 624.
(c) "Phenothiazines", "Chemistry of Carbon Compounds", ed. E. H. Rodd, Vol. IV C, p. 1512.

**References**

1. E. Knoevenagel, *J. prakt. Chem.*, 1914, **89**, 1.
2. D. F. Houston, E. B. Kester, and F. De Eds, *J. Amer. Chem. Soc.*, 1949, **71**, 3816.

## 2-OXO-1,4-BENZODIOXAN

Catechol condenses with chloroacetyl chloride in the presence of base to yield 2-oxo-1,4-benzodioxan (I).[1] The reaction probably involves initial acylation of the phenol, followed by intramolecular alkylation.

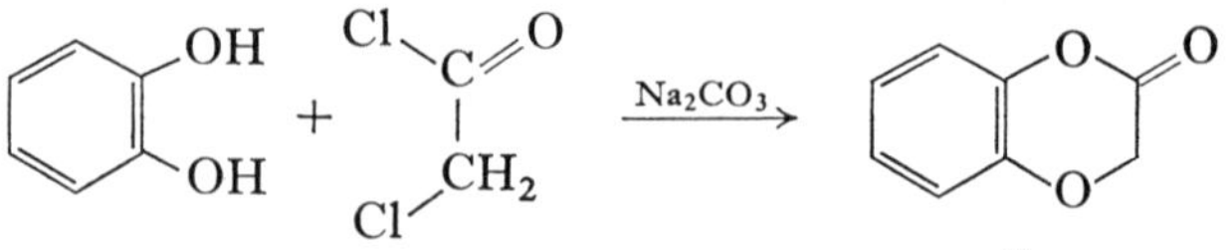


I

### 2-Oxo-1,4-benzodioxan (I)

To a solution of catechol (11 g.) in acetone (450 ml.), contained in a 1-litre 3-necked flask, equipped with reflux condenser, dropping-funnel and mercury-sealed stirrer, add anhydrous sodium carbonate (10·9 g.). Heat the mixture under reflux on a water-bath at 100°, and add dropwise over 20 min. a solution of chloroacetyl chloride (11·3 g.) in acetone (50 ml.). Continue heating under reflux overnight with vigorous stirring. Cool the mixture, filter off the inorganic material and evaporate off the excess of solvent. Dissolve the oily residue in ether (100 ml.), transfer the ethereal solution into a separating funnel, and wash successively with 0·5N sodium hydroxide (2 × 30 ml.) and water (2 × 30 ml.). Dry the ethereal solution ($MgSO_4$), and evaporate off the ether to obtain the product as a sweet-smelling oil that solidifies on cooling.

2-Oxo-1,4-benzodioxan crystallizes (with difficulty) from ether (ca. 4 ml./g.) as pale-yellow prisms, m.p. 53–54°.

Yield = 7–8 g. (46·5–53·5%).

**Review**

"Benzo-1,4-dioxanes", "Heterocyclic Compounds", ed. R. C. Elderfield, Vol. 6, p. 59.

**Reference**

1. T. Bingham and R. K. Smalley, University of Salford, unpublished results.

## PHENOXATHIIN

Phenoxathiin (II) is readily prepared by heating diphenyl ether (I) with sulphur at 100° in the presence of aluminium chloride.[1]

### Phenoxathiin (II)

**Note**

It is advisable to carry out this preparation in an efficient fume cupboard, since hydrogen sulphide is evolved during the reaction.

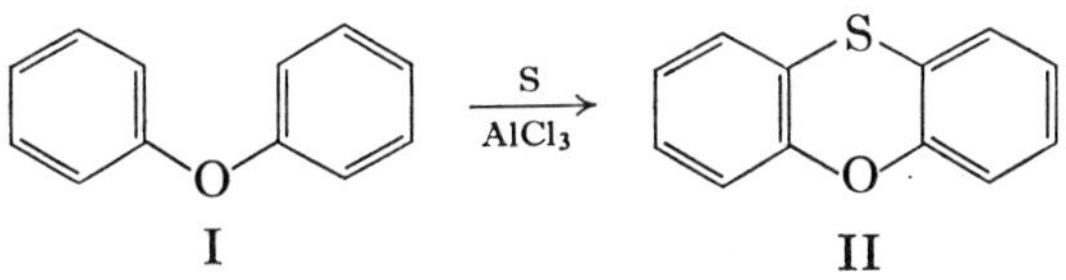

In a 100-ml. round-bottomed flask prepare an intimate mixture of diphenyl ether (17 g.; 15·85 ml.), flowers of sulphur (2·6 g.) and anhydrous aluminium chloride (5·1 g.). Equip the flask with a reflux condenser, and heat the mixture on a water-bath at 100° until hydrogen sulphide evolution ceases (ca. 4 hr.). Cautiously stir the molten mixture into a solution of concentrated hydrochloric acid (10 ml.) in ice–water (100 ml.), then transfer the resulting solution into a separating funnel and extract with ether (3 × 40 ml.). Combine the extracts, dry ($MgSO_4$), and after evaporating off the ether, distil the residual oil under oil-pump vacuum. Neglect the fraction b.p. 90–150°/5 mm., which is mainly diphenyl ether. The product is obtained as a colourless oil, b.p. 180–185°/5 mm., which solidifies on cooling.

Phenoxathiin crystallizes from methanol (ca. 3 ml./g.) as colourless plates, m.p. 56–57°.

Yield = 5–6 g. (25–30%).

## Reviews

(a) "The Chemistry of Phenoxathiin and its Derivatives", C. L. Deasy, *Chem Rev.*, 1943, **32**, 173.

(b) "Sulphur analogs of the Dioxanes", "Heterocyclic Compounds", ed. R. C. Elderfield, Vol. 6, p. 81.

## Reference

1. C. M. Suter and C. E. Maxwell, *Org. Synth.*, 1943, Coll. Vol. II, 485.

# 9 | MISCELLANEOUS SYSTEMS

## CYCLOHEXA-2,5-DIENE-4-ONE-SPIRO-2′-OXETANO-[3′,4′]CYCLO-OCTANE

Irradiation of a mixture of *p*-benzoquinone and *cis*-cyclo-octene in benzene solution leads to photoaddition of the quinone to the olefin yielding cyclohexa-2,5-diene-4-one-spiro-2′-oxetano[3′,4′]cyclo-octane (I).[1]

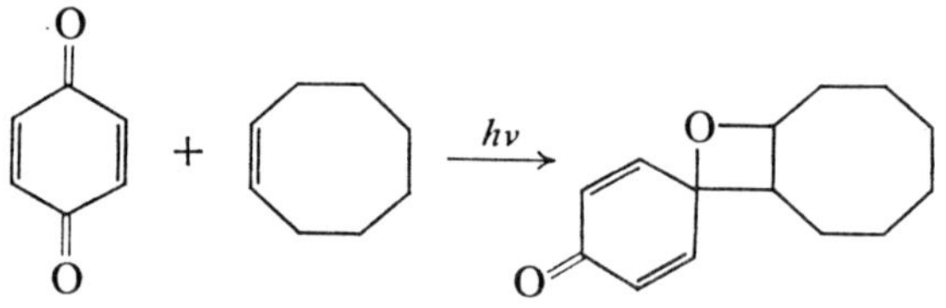


I

### Cyclohexa-2,5-diene-4-onespiro-2′-oxetano[3′,4′]cyclo-octane (I)

In a water-cooled borosilicate glass tube, dissolve *p*-benzoquinone (2 g.) in a mixture of *cis*-cyclo-octene (20 ml.) and dry benzene (100 ml.). Irradiate the solution with a 500-watt medium-pressure mercury arc lamp† for 3–4 hr., after which time the quinone absorption in the 430–450 nm region will have disappeared. Evaporate off the solvents under reduced pressure and distil the residual oil under oil-pump vacuum.

† In place of the medium-pressure mercury arc lamp, ordinary tungsten light bulbs or sunlight can be used quite successfully, but the irradiation time is appreciably longer.

Cyclohexa-2,5-diene-4-onespiro-2′-oxetano[3′,4′]cyclo-octane is obtained as a colourless oil, b.p. 120–130°/0·2 mm.

Yield = 3·4–3·8 g. (84–94%).

**Spectra**

(a) Ultraviolet.[1]
(b) Infrared.[1]
(c) Proton magnetic resonance.[1]

**Reviews**

(a) "Oxetanes", "The Chemistry of Heterocyclic Compounds", ed. A. Weissberger, Vol. 19, p. 983.
(b) "Trimethylene Oxides", "Heterocyclic Compounds", ed. R. C. Elderfield, Vol. 1, p. 59.
(c) "Trimethylene Oxides (Oxetans)", "Chemistry of Carbon Compounds", ed. E. H. Rodd, Vol. IV A, p. 20.
(d) N. J. Turro, "Molecular Photochemistry", W. A. Benjamin, Inc., New York, 1965, p. 208.
(e) R. O. Kan, "Organic Photochemistry", McGraw-Hill Book Co., New York, 1966, p. 198.

**Reference**

1. D. Bryce-Smith, A. Gilbert, and M. G. Johnson, *J. Chem. Soc.*, 1967, 383.

## 4,6-DINITROBENZIMIDAZO[1,2-*a*]PYRIDINE

2-Aminopyridine condenses with picryl chloride (I) in benzene solution to give 2-(2′,4′,6′-trinitroanilino)pyridine (II), which on heating in dimethylaniline solution cyclizes with loss of nitrous acid, to 4,6-dinitrobenzimidazo[1,2-*a*]-pyridine (III).[1]

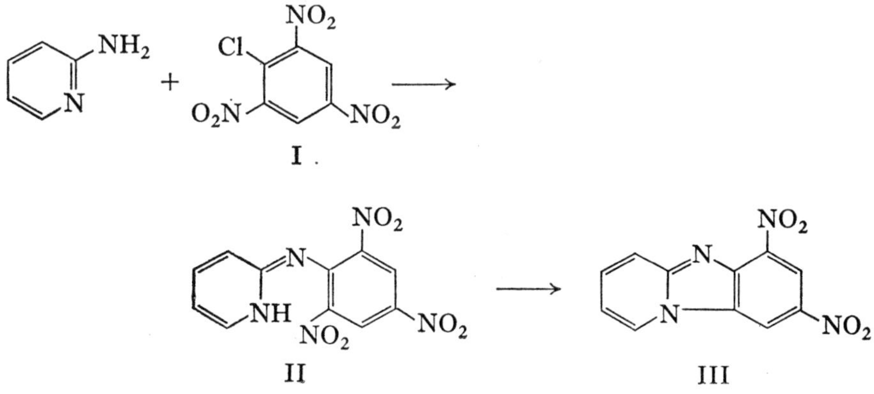

### 2-(2′,4′,6′-Trinitroanilino)pyridine (II)

To a hot solution of 2-aminopyridine (5·7 g.) in benzene (50 ml.), contained in a 250-ml. 3-necked, round-bottomed flask, fitted with reflux condenser and dropping-funnel, add over a period of 15 min. a solution of picryl chloride (7·4 g.) in benzene (50 ml.). Heat the resulting solution under reflux on a water-bath at 100° for 1 hr. Filter the hot mixture, evaporate off the solvent, and stir the residual oil with petroleum ether (b.p. 40–60°; 50 ml.) until solidification occurs. Filter off the product and dry at 100°.

2-(2′,4′,6′-Trinitroanilino)pyridine crystallizes from benzene (ca. 6 ml./g.) as yellow rhombs, m.p. 135–136°.

Yield = 7·8–8·2 g. (63·5–66·5%).

### 4,6-Dinitrobenzimidazo[1,2-*a*]pyridine (III)

In a 100-ml. round-bottomed flask, heat a solution of 2-(2′,4′,6′-trinitroanilino)pyridine (5 g.) in dimethylaniline (10 ml.) under reflux for 2 hr. Cool the solution, filter off the crude product, and wash successively with cold ethanol (2 × 30 ml.) and petroleum ether (b.p. 40–60°; 2 × 40 ml.).

4,6-Dinitrobenzimidazo[1,2-*a*]pyridine crystallizes from nitrobenzene (ca. 5 ml./g.) (decolourizing carbon) as orange needles, m.p. >320°.

Yield = 3·2–3·6 g. (80–88%).

**Review**

"Pyrido[1,2-*a*]benzimidazoles", "The Chemistry of Heterocyclic Compounds", ed. A. Weissberger, Vol. 15 (part 1), p. 507.

**Reference**

1. G. Morgan and J. Stewart, *J. Chem. Soc.*, 1938, 1292.

## 2-PHENYLIMIDAZO[1,2-*a*]PYRIDINE

2-Aminopyridines are alkylated on the ring nitrogen atom by phenacyl halides,[1,2] and the products cyclize on heating to yield imidazo[1,2-*a*]pyridines. Interaction of 2-aminopyridine (I) with phenacyl bromide in the presence of sodium hydrogen carbonate yields 2-phenylimidazo[1,2-*a*]pyridine (II).[3,4]

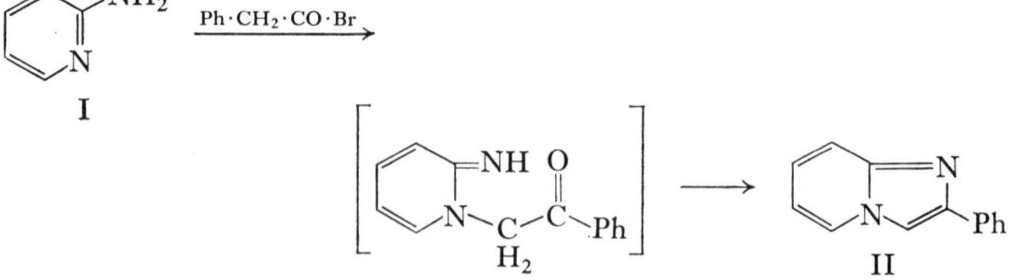

### 2-Phenylimidazo[1,2-*a*]pyridine (II)

To a solution of 2-aminopyridine (5·8 g.) in ethanol (60 ml.) contained in a 250-ml. round-bottomed flask, add phenacyl bromide (10 g.), and sodium hydrogen carbonate (8·4 g.). Stir the resulting mixture at room temperature for 2 hr., and then heat under reflux on a water-bath at 100° for a further 2 hr. Cool the mixture, pour into cold water (200 ml.), and stir until solidification occurs. Filter off the crude product, and wash with cold water (5 × 60 ml.).

2-Phenylimidazo[1,2-*a*]pyridine crystallizes from 60% aqueous ethanol (ca. 8 ml./g.) as buff prisms, m.p. 132–134°.

Yield = 7·5–8 g. (62·5–67%).

#### Spectra

Ultraviolet.[5]

#### Reviews

(a) B. Kickhöfen, *Arch. Pharmazie*, 1955, **288**, 473.
(b) "Imidazo[1,2-*a*]pyridines", "The Chemistry of Heterocyclic Compounds", ed. A. Weissberger, Vol. 15 (part 1), p. 460.

#### References

1. R. Delaby, J. Mandereau, and P. Reynaud, *Bull. Soc. chim. France*, 1961, 2065.
2. A. P. Gray, and D. E. Heitmeier, *J. Amer. Chem. Soc.*, 1959, **81**, 4347.
3. I. A. Kaye, C. L. Parris, and W. J. Burlant, *J. Amer. Chem. Soc.*, 1953, **75**, 746.
4. P. L. Julian, E. W. Meyer, A. Magnini, and W. Cole, *J. Amer. Chem. Soc.*, 1945, **67**, 1203.
5. J. D. Bower, *J. Chem. Soc.*, 1957, 4510.

## 2,3,4,5-TETRAHYDROPYRIDO[1,2-*a*]BENZIMIDAZOLE

Oxidative cyclization of *N*-(*o*-formamidophenyl)piperidine with performic acid gives an excellent yield of 2,3,4,5-tetrahydropyrido[1,2-*a*]benzimidazole[1] (III).[2,3] The cyclization has been shown[2] to proceed via a mechanism related to that of the Polonovski reaction.[4] The required *N*-(*o*-aminophenyl)piperidine (II) is readily prepared by condensing piperidine with *o*-nitrochlorobenzene, and reducing the resulting nitro-compound (I) with stannous chloride and hydrochloric acid.

### *N*-(*o*-Nitrophenyl)piperidine (I)

Add *o*-nitrochlorobenzene (7·9 g.) to piperidine (25 ml.) contained in a 100-ml. round-bottomed flask, and heat the mixture on a water-bath at 100° for 1·5 hr.

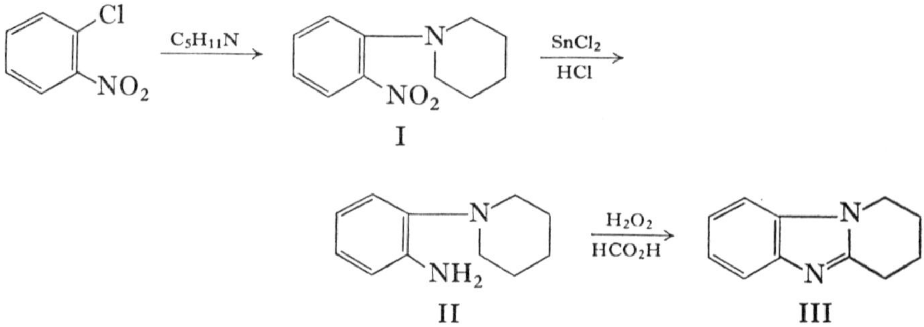


Pour the cooled solution into cold water (200 ml.), and stir vigorously until the product solidifies. Filter off the product and wash with cold water (4 × 60 ml.).

*N*-(*o*-Nitrophenyl)piperidine crystallizes from ethanol (ca. 5 ml./g.) as orange prisms, m.p. 88–89°.

Yield = 8·1–8·8 g. (79–85·5%).

### *N*-(*o*-Aminophenyl)piperidine (II)

To a hot solution of stannous chloride (20 g.) in 6N hydrochloric acid (75 ml.) contained in a 250-ml. 2-necked, round-bottomed flask, add dropwise over a period of 1 hr. a solution of *N*-(*o*-nitrophenyl)piperidine (5 g.) in 6N hydrochloric acid (30 ml.). Heat the resulting solution under reflux for 1 hr., and then cautiously pour the hot mixture into a 1-litre round-bottomed flask, containing an ice-cold solution of sodium hydroxide (100 g.) in water (200 ml.). Steam distil the mixture for ca. 3 hr., and filter off the solid product from the distillate.

*N*-(*o*-Aminophenyl)piperidine is obtained as a white solid, m.p. 44–45°, sufficiently pure for direct use in the next stage.

Yield = 3·5–4 g. (79–91%).

### 2,3,4,5-Tetrahydropyrido[1,2-*a*]benzimidazole (III)

Warm a solution of *N*-(*o*-aminophenyl)piperidine (2 g.) in 98% formic acid (12 ml.) on a water-bath at 100° for 5 min., then cool, and add 30% hydrogen peroxide (6 ml.). Cautiously heat the solution on a water-bath at 100°, and when the vigorous exothermic reaction has subsided (ca. 20 min.), cool, and pour the reaction mixture into cold water (100 ml.). Basify the solution by adding an excess of solid sodium carbonate, and filter off the product.

2,3,4,5-Tetrahydropyrido[1,2-*a*]benzimidazole crystallizes from petroleum ether (b.p. 80–100°; ca. 8 ml./g.) as colourless prisms, m.p. 101–102°.

Yield = 1·35–1·55 g. (72–85·5%).

**Spectra**

Ultraviolet.[2]

## Reviews

"Benzimidazoles", see review references on p. 43.

## References

1. K. H. Saunders, *J. Chem. Soc.*, 1955, 3275.
2. O. Meth-Cohn and H. Suschitzky, *J. Chem. Soc.*, 1963, 4666.
3. R. Adams and M. D. Nair, *J. Amer. Chem. Soc.*, 1961, **83**, 3518.
4. M. Polonovski and M. Polonovski, *Bull. Soc. chim. France*, 1927, **41**, 1190.

## 1,4-DIHYDRO-*endo*-1,4-OXANAPHTHALENE

In certain reactions, furan (I) exhibits the properties of a conjugated diene. It will, for example, undergo a Diels–Alder[1] addition reaction with a suitable dienophile, such as benzyne (II). In this example benzyne is generated *in situ* by the oxidation of 1-aminobenzotriazole (III) (see preparation on p. 45) with lead tetra-acetate.[2]

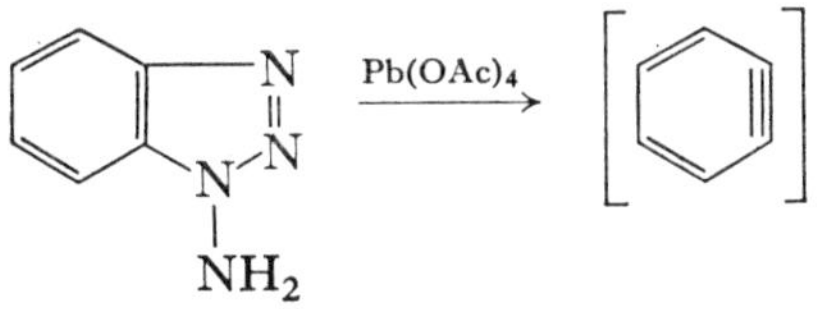


III II

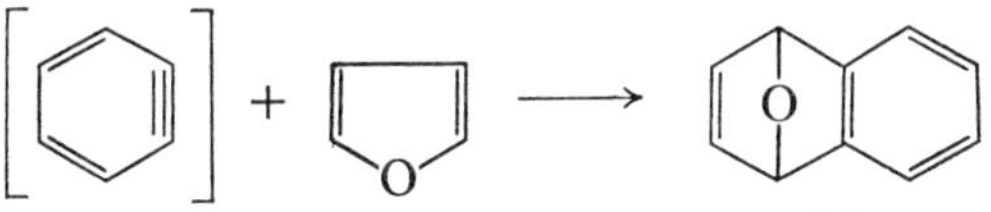


II I IV

### 1,4-Dihydro-*endo*-1,4-oxanaphthalene (IV)

To a magnetically stirred suspension of lead tetra-acetate (10 g.) in dry furan (20 ml.) contained in a 100 ml. conical flask, add dropwise over 5 min. a solution of 1-aminobenzotriazole (1·34 g.) in dry furan (40 ml.).

After the vigorous effervescence has ceased, stir the mixture at room temperature for 5 min.; then filter and evaporate the filtrate to dryness. To the residue add ether (25 ml.), stir, filter to remove the insoluble material, and evaporate to dryness. Repeat the procedure once more to obtain the crude product as a brown oil. Extract this oil with hot petroleum ether (b.p. 60–80; 2 × 25 ml.), combine the extracts and cool in an ice-bath.

1,4-Dihydro-*endo*-1,4-oxanaphthalene crystallizes slowly and with difficulty from the petroleum ether extracts as large colourless prisms, m.p. 54–56°.

Yield = 0·7–0·9 g. (50–62·5%).

A higher yield (75–80%) of product may be obtained by careful vacuum distillation of the crude product before crystallization.

### Reviews

(a) "The Diels–Alder Diene Synthesis", J. A. Norton, *Chem. Rev.*, 1942, **31**, 319.
(b) "Diels–Alder Synthesis with Hetero-aromatic Compounds", S. B. Needleman and M. C. Chang-Kuo, *Chem. Rev.*, 1962, **62**, 405.
(c) "Diels–Alder Reactions Part I. New Preparative Aspects", J. Sauer, *Angew. Chem.* (*Internat. edn.*), 1966, **5**, 211.
(d) "Diene Synthesis with Dienophiles containing Hetero-atoms", Yu. A. Arbuzov, *Russ. Chem. Rev.*, 1964, **33**, 407.
(e) "The Chemistry of Benzyne", J. F. Bunnett, *J. Chem. Educ.*, 1961, **38**, 278.
(f) "Benzyne and Related Intermediates", H. Heaney, *Chem. Rev.*, 1962, **62**, 81.

### References

1. O. Diels, K. Alder, and E. Naujoks, *Ber.*, 1929, **62**, 554.
2. C. D. Campbell and C. W. Rees, University of Leicester, private communication.

# BIBLIOGRAPHY

"The Chemistry of Heterocyclic Compounds", ed. A. Weissberger. Interscience Publishers, New York, 1950–1967, Vols. 1–23.

"Heterocyclic Compounds", ed. R. C. Elderfield. John Wiley & Sons Inc., New York, 1950–1967, Vols. 1–8.

"Chemistry of Carbon Compounds", ed. E. H. Rodd. Elsevier Publishing Company, Amsterdam, 1957–1960, Vols. IVA–IVC.

"Advances in Heterocyclic Chemistry", ed. A. R. Katritzky and A. J. Boulton. Academic Press, New York and London, 1963–1967, Vols. 1–8.

"The Principles of Heterocyclic Chemistry", A. R. Katritzky and J. M. Lagowski. Methuen and Company Ltd, London, 1967.

"Introduction to the Chemistry of Heterocyclic Compounds", R. M. Acheson. John Wiley & Sons Inc., London, 1967, 2nd Edn.

"A Text-book of Practical Organic Chemistry including Qualitative Organic Analysis", A. I. Vogel. Longmans, Green and Company, London, New York and Toronto, 1956, 3rd Edn.

# SUBJECT INDEX

## C

## D

## E



## F



## G

## H



## I



## J



## K



## L



## M

## Q



## R